Helium and Hydrogen Isotope Adsorption and Separation in Metal-Organic Frameworks

Helium and Hydrogen Isotope Adsorption and Separation in Metal-Organic Frameworks

Von der Fakultät Mathematik und Physik
der Universität Stuttgart
zur Erlangung der Würde eines Doktors der
Naturwissenschaften (Dr. rer. nat.) genehmigte Abhandlung

Vorgelegt von

Ingrid Zaiser (geb. Weinrauch)
aus Valenii de Munte (Rumänien)

Hauptberichterin: Prof. Dr. G. Schütz

Mitberichter: Prof. Dr. C. Bechinger

Eingereicht am: 25.08.2016

Tag der mündlichen Prüfung: 26.09.2016

Max-Planck-Institut für Intelligente Systeme, Stuttgart

2016

Bibliografische Information der Deutschen Nationalbibliothek
Die Deutsche Nationalbibliothek verzeichnet diese Publikation in der Deutschen Nationalbibliografie; detaillierte bibliografische Daten sind im Internet über http://dnb.d-nb.de abrufbar.

1. Aufl. - Göttingen: Cuvillier, 2016
Zugl.: Stuttgart, Univ., Diss., 2016

Nonnenstieg 8, 37075 Göttingen
Telefon: 0551-54724-0
Telefax: 0551-54724-21
www.cuvillier.de

1. Auflage, 2016
Gedruckt auf umweltfreundlichem, säurefreiem Papier aus nachhaltiger Forstwirtschaft.

ISBN 978-3-7369-9377-8
eISBN 978-3-7369-8377-9

Contents

1 Introduction

Hydrogen and helium are both lightest and most abundant elements in the universe. Still, the extraction of their rare, but highly demanded isotopes like deuterium and helium-3 is a challenging task. Deuterium occurs naturally in a low abundance of only 0.015% [1] and helium-3 is extremely rare on earth with a natural abundance of 0.00013% [2]. Many industrial and scientific applications require deuterium and cannot at all be performed with a mixture of hydrogen isotopes. Deuterium is needed for example in research reactors as a cold source, in neutron scattering and for isotopic tracing during chemical reactions. Helium-3 on the other hand is highly requested for use in neutron detectors [3] or in dilution refrigerators [4, 5] to reach extremely low temperatures of millidegrees Kelvin. Even the possibility of lunar mining [6, 7] has been taken into consideration to satisfy the demand of this precious isotope. Deuterium and helium-3 can be used as fuel for nuclear fusion. Operating with deuterium-tritium fuel for fusion is useful for studies, but less practical for commercial application. This reaction produces large amounts of radiation in form of neutrons and a substitution of the radioactive tritium with helium-3 reduces the neutron production. Recently, a breakthrough has been reported in the design of the stellarator concept Wendelstein-7X. [8] The chance of realizing nuclear fusion as future energy supply supports the great importance to extract these isotopes from their most common isotope form, i.e. hydrogen and helium-4.

A widespread technique for the extraction of molecules from gas mixtures uses shape selective microporous materials, called molecular sieves. This technique is not applicable for isotopes, as different isotopes of the same element have identical size. Nowadays deuterium is extracted via cryogenic distillation at 24 K or the Girdler-sulfide process. Both methods have very low separation factors (1.5 and 2.3, respectively [9]) and consume high amounts of energy. The standard way to produce helium-3 is coupled with the production of tritium in the nuclear weapon industry. Helium-3 is skimmed as a byproduct of the radioactive tritium decay. Direct recovery due to the occurrence of traces of helium-3 in the atmosphere and in some natural gas sources is also possible.

The high energy consumption and low efficiency of conventional isotope separation methods, which are based on mass differences, motivates to search for alternatives. Quantum Sieving has been proposed by Beenakker *et al.* [10] as a promising technique for the separation of light isotopes, which are selectively adsorbed in very small pores at low temperatures. The idea is based on the favorable adsorption of the heavier isotope with the lower zero-point energy. This isotope is preferred to enter the pores. Another concept for the separation of isotopes, called Chemical Affinity Sieving uses as well the difference in zero-point energy. [11] When the isotopes interact with strongly attractive adsorption centers, this difference can be enhanced and results in higher adsorption enthalpy of the heavier isotope and, therefore, separation.

The new class of crystalline porous materials, called metal-organic frameworks (MOFs) provides excellent candidates for the separation of isotopes by either Quantum Sieving or Chemical Affinity Sieving. MOFs are assembled by metal-ions as framework nodes which are connected via organic ligands as linkers. The great variety of the building units leads to different pore sizes and possible functionality. For Quantum Sieving, MOFs with suitable pore sizes are demanded and for Chemical Affinity Sieving, strong adsorption sites like open metal sites are required.

The separation of light isotopes is currently a topic of high interest in the scientific community. Recently, the group of the Nobel Prize holder Prof. Geim has published the separation of hydrogen isotopes with graphene and boron nitride sieves at room-temperature. [12] The greater zero-point energy of protons allows them to overcome the materials energy barriers easier. Therefore, deuterons permeate through this crystal much slower than protons, which results in a separation factor of approximately 10. This is a great result at room temperature, but from a practical point of view a preference of the less abundant deuteron over the proton would be better. In the reported case a lot of energy is spent to transport the much larger amount of protons across the sieve. Also helium isotope separation with graphene is currently investigated. Theoreticians proposed the separation of helium isotopes with nanoporous functionalized graphene membranes. [13, 14] They removed two neighbored rings of carbon in the graphene membranes and saturated the carbon atoms with hydrogen and nitrogen atoms in order to create a suitable barrier height, which leads to a higher transmission probability for helium-3. A helium-3/helium-4 ratio of 19 has been calculated for gas at 10 K, which decreases very fast to a ratio of 2 at 20 K. For testing this concept experimentally, the creation of suitable nanosized holes in graphene is still demanded.

The separation of hydrogen and helium isotopes in MOFs has been experimentally investigated with thermal desorption spectroscopy (TDS) in the present work. TDS gives a fingerprint of the adsorbed isotopes in the sample. After the material has been exposed to equimolar isotope mixtures the separation factor, called selectivity, can be directly determined from TDS spectra. The adsorbed amount of gas is proportional to the area enclosed by the TDS maxima.

Several MOFs with pore apertures meeting the size requirement for separation of hydrogen and helium isotopes with Quantum Sieving have been chosen. Adequate candidates have been found in the IFP [15–17] and MFU-4 [18] MOF series. MOFs with open metal sites are used in this work for the investigation of Chemical Affinity Sieving. One material with a very strong open copper site, named Cu(I)-MFU-4l [19] and three open nickel site materials, CPO-27 [20,21], DUT-9 [22] and CFA-6 [23] have been investigated.

2 Fundamentals

2.1 Adsorption in porous materials

Adsorption occurs when particles of a gas or a liquid stick to the surface of a solid material (adsorbent). In contrast to the bulk atoms of a solid, the atoms on the surface are not wholly surrounded by other atoms and therefore the bonds on the surface are unsaturated. In order to reach an energetically more favorable state the surface atoms attract the gas molecule which looses through adsorption at least one degree of freedom of translation and thus the entropy (ΔS) decreases. For any spontaneous process the Gibbs free energy decreases ($\Delta G < 0$). In accordance with the thermodynamic equation

$$\Delta G = \Delta H - T\Delta S \tag{2.1}$$

the enthalpy ΔH also decreases. The nature of the involved forces between the adsorbent and the gas molecules are of two main kinds, physical and chemical. Sharing or transfer of electrons between adsorbent and gas molecules leads to chemisorption of the gas molecule, which can dissociate on the surface. In case of chemisorption only a monolayer can be adsorbed on the surface. The chemisorption bond has the characteristics of a chemical bond and is site specific i.e. localized at active centres on the adsorbent.
The physical interaction (physisorbtion), is an exothermic process and leads to a reversible bound mono- or multilayer of gas molecules. [24] Adsorption through physisorption arises from van der Waals forces. Fluctuations in the charge distribution of polarizable molecules induces a temporary dipole, which affects the surrounding molecules by inducing a corresponding electrical moment in a near neighbor molecule leading to attractive interactions (London-type dispersion). The counterpart is the short-range repulsive interaction arising from electron orbital overlap (Pauli exclusion principle). Both forces get balanced when adsorption occurs. The Lennard-Jones potential describes the interaction between a gas molecule and an adsorbent surface in dependence of their distance r [25]:

$$V_{LJ} = 4\epsilon\left[\left(\frac{\sigma}{r}\right)^{12} - \left(\frac{\sigma}{r}\right)^{6}\right] \tag{2.2}$$

ϵ is the depth of the potential well. σ is known as the van der Waals radius and represents the smallest distance between two non-bonding particles. The r^{-6} term of the Lennard-Jones potential describes the attractive long-range interactions and has been first proposed by F. London [26] in 1930. The repulsive part has a r^{-12} dependence and is in contrast to the London dispersion interaction not theoretically justified. The Pauli repulsion is well approximated by this dependence and is quiet convenient for calculations, since it is the square of the attractive term. This potential is also valid for the adsorption of single particles on a flat surface. Figure 2.1a shows the Lennard-Jones potential including the relevant parameters.

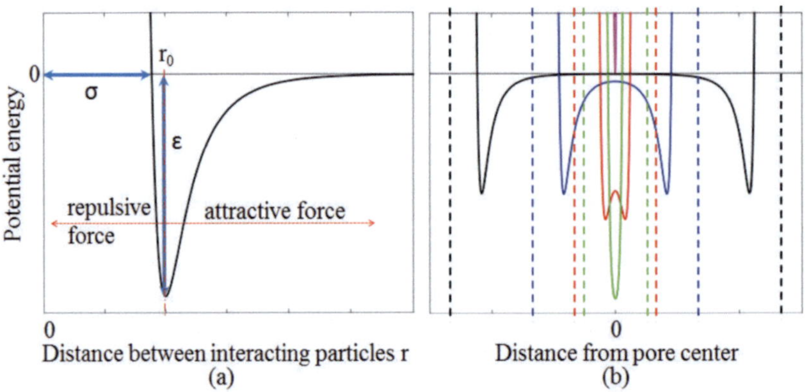

Figure 2.1: a) Lennard-Jones potential for a particle approaching a surface b) Interaction potential for a particle in slit-pores with different distances, indicated by different colors. Ratio between pore and particle diameter: 10, 5, 2.5, 2, 1. The dashed line represents the pore wall.

The adsorption in porous materials can be described in the simplest model by two flat surfaces interacting on the particle, the so-called slit pore. [27] Figure 2.1b depicts the potential of the particle in the slit pore as a function of the distance between the two surfaces for different ratios of pore and particle diameter. Compared to a single flat surface, the potential energy exhibits two minima near each pore wall. These minima increase in magnitude and coalesce into one minimum for smaller pore diameters. The repulsive parts overlap for pore to particle size ratios near 1 and the potential depth decreases rapidly. In contrast to a flat surface, where the adsorbed particle interacts only with one or two surface atoms, a curved surface leads to an overlap of the van der Waals potentials. This overlap increases the adsorption strength for the particle even more for smaller pore diameters. [28–31] Additionally, the surface of porous materials may be composed of different elements, which can add an energetic heterogeneity to the adsorption potential.

2.1.1 Dihydrogen complexes

Dihydrogen bond to metal cations form complexes M^+-H_2, which provide a bridge between physisorbed and chemisorbed, i.e. between bound and dissociated states of dihydrogen. H_2 is bond to the metal center, but the H-H bond is still present. This bonding mechanism has not been found until 1984. [32]

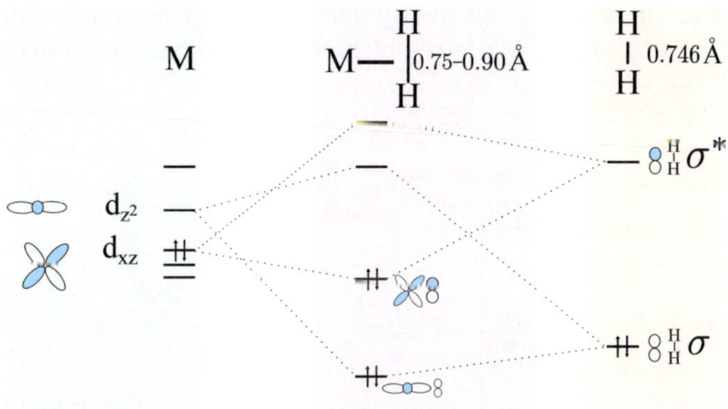

Figure 2.2: Illustration of the formation of dihydrogen complexes. The metal with its corresponding d orbitals is presented on the left side, the bonding (σ) and antibonding (σ^*) orbitals of hydrogen on the right side. In the middle the formation of the dihydrogen metal complex is shown. The bond length of the gaseous H_2 and the range for the elongation of the bond length due to the building of the complex are given above the energy levels. [33]

For building the M^+-H_2 complex, H_2 donates an electron from its σ bonding orbital into the d-orbitals of a metal and a back-donation from the metal occurs in the empty σ^* antibonding orbital (figure 2.2). Intuitively, an antibonding orbital forming a chemical bond seems contradictory, but the σ^* orbital is only antibonding with respect to the bond between the H atoms and can still be bonding for the metal dihydrogen complex. The electron in the σ^* antibonding orbital and the half filled σ bonding orbital lead to a weakening of the H-H bond. Therefore, the bond length increases from 0.746 Å of a gaseous H_2 to values typically in the range of 0.75 Å-0.90 Å [34]. This increase in bond length comes with a decrease in vibrational frequency from 4160 cm^{-1} to a range between 2600-3000 cm^{-1} [35].

2.2 Isotope separation methods

2.2.1 Current methods for industrial separation of hydrogen isotopes and extraction of helium-3

Due to their identical chemical properties, hydrogen isotopes can only be separated by methods which use their different mass. The standard method for the enrichment of heavy water is the bithermal Girdler-sulfide process.

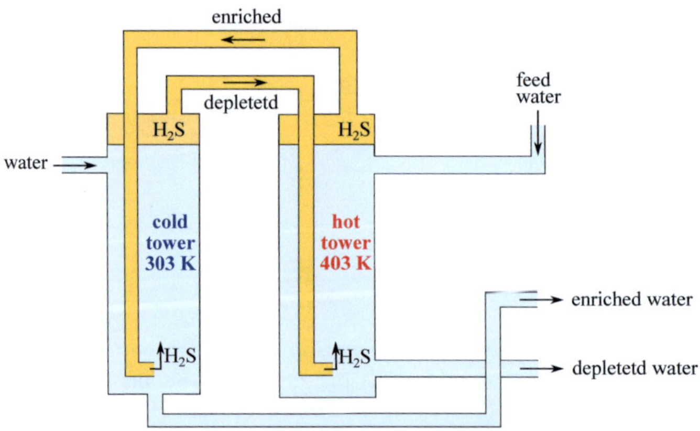

$$H_2O(l) + HDS(g) \rightleftharpoons HDO(l) + H_2S(g)$$

Figure 2.3: Concept of the Girdler-sulfide process for the production of heavy water.

This process uses the different equilibria of the reaction presented in figure 2.3 at 303 K and 403 K. [36] Hydrogen sulfide circulates as a gas in a closed loop between a cold tower and a hot tower, both filled with water. In the cold tower deuterium atoms migrate preferentially from the hydrogen sulfide gas to the liquid water. Fresh normal water with the natural abundance of deuterium is filled to the hot tower. There, the deuterium migrates from the water to the hydrogen sulfide gas, thereby enriching this gas. This enriched gas then again flows through the cold tower and further enrichment of the water takes place. The water leaving the hot tower has a lower concentration of deuterium atoms compared to when it entered the system. Multiple processing cycles lead to enriched water with 20%-30% [36] deuterium content. Further increasing of the purification is done with a vacuum distillation system, leading to a purity of approximately 99.75% of deuterium. The overall separation factor of this process is only 2.3 [9].

Another widespread technique for the purification of gases is the cryogenic distillation. The gases are cooled until they are liquefied and afterwards the components are selectively distilled at their individual boiling temperatures. Cooling to the low boiling points of hydrogen isotopes (20 K and 24 K [37] for H_2 and D_2, respectively) requires high energy consumption for the cooling process and their small temperature difference leads to low separation factor of 1.5 at 24 K [9].

Helium-3 is usually extracted as a byproduct of the radioactive tritium decay from tritium reserves of the nuclear weapons programs in the U.S. and Russia. [38] The reduction of the nuclear weapons stockpile and the increased demand for several applications resulted in a shortage of helium-3. This isotope also occurs in the atmosphere or in some natural gas sources. Therefore, direct extraction is also an option but was not used extensively in the past, as the supply with helium-3 was greater than the demand. [39]

2.2.2 Isotope separation with Quantum Sieving

In 1995 Beenakker *et al.* developed the theory of Quantum Sieving while investigating the adsorption of hard sphere molecules in cylindrical pores. When the difference between pore diameter d and hard core diameter σ of the molecule, becomes comparable to the de Broglie wavelength of the molecule [10]

$$\lambda_{deBroglie} \approx d - \sigma \tag{2.3}$$

a quantum mechanical treatment is required. This quantum mechanical treatment is particularly necessary at low temperature, where the de Broglie wavelength is larger. The model has been developed under the assumption of no molecule-molecule interaction and no molecule-surface interaction stronger than the van der Waals type. Therefore, the motion of the molecules in the pores is separated into the independent axial and radial ones. The axial motion is free between two consecutive collisions with phonons or surface defects, but the radial direction is restricted by a potential well created by the interaction with the pore walls. This potential has been approximated by a circular square well with the depth ϵ with respect to zero potential energy for a molecule in free space (see figure 2.4 left).

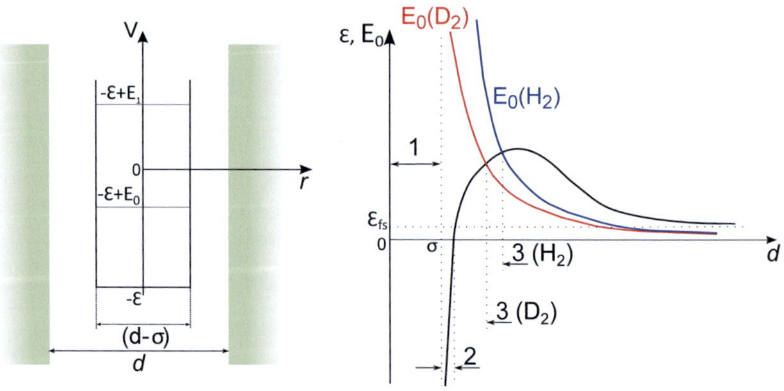

Figure 2.4: Left: Representation of the potential energy, V, for a molecule confined in a pore with $d - \sigma \approx \lambda$ and the transverse motion energy levels E_i. Right: The potential well depth ϵ and the zero-point energies of the isotopes in dependence on the pore diameter.

The energy levels E_i for a circular square well are [40]:

$$E_i = \frac{2\gamma_i^2\hbar^2}{(d-\sigma)^2 m}, \tag{2.4}$$

where γ_i are related to the zeros of the Bessel functions. The potential well depth ϵ which represents the attraction of the walls and the zero-point energies are illustrated in figure 2.4 right as functions of the pore diameter. No adsorption is possible in region 1 because the hard core diameter exceeds the pore diameter. In region 2, where the pore gets slightly bigger than the hard core diameter, penetration is still difficult because the zero-point motion overcompensates the attraction of the molecule by the pore surface. Thus, the molecules encounter an energy barrier and the process of entering into the channel requires thermal excitation. In case of large pores the interaction potential is not affected by the opposite wall or curvatures and ϵ is simply the flat surface potential. The adsorption of a system is characterized by the relation of the zero-point energy and the potential well depth. The potential well is attractive for the molecule as long as the attraction of the walls overcompensates the zero-point energy of the molecule ($\epsilon > E_0$). Figure 2.4 right depicts the different zero-point energies for hydrogen and deuterium according to their different masses. Hydrogen and deuterium have a mass ratio of 1:2, which is the largest value known for any two isotopes. With decreasing pore diameter hydrogen reaches earlier (at bigger diameters) the point marked with number 3, where it becomes more difficult to enter the pore compared to deuterium. The region where the zero-point energy of hydrogen already exceeded the attraction of the walls and penetration of deuterium is still possible represents the Quantum Sieving domain. As a consequence, in small pores and at temperatures below 100 K deuterium diffuses into the material faster than the lighter hydrogen. [41]

2.2.3 Isotope separation with Chemical Affinity Sieving

Chemical Affinity Sieving is a general concept for gas separation and the most commonly observed phenomenon in molecular sieving. [42] Molecules, which have the strongest chemical affinity with the pores are predominantly adsorbed by the pores.

A hydrogen molecule in the gas-phase has six degrees of freedom, namely three translational, two rotational and one vibrational, which contribute to the zero-point energy. If the adsorption potential is high, a barrier for diffusion on the surface exists and the molecule looses the translational degrees of freedom parallel to the surface. The translational motion of the adsorbing molecule towards the surface can be thought of as the vibration of the species against the surface, i.e., to a vibrational motion in the potential energy well, which defines the adsorption itself. [43] Thus, the strength of an adsorption site has an impact on the zero-point energy of the adsorbed isotope. The different mass of isotopes leads to different zero-point energies for each isotope.

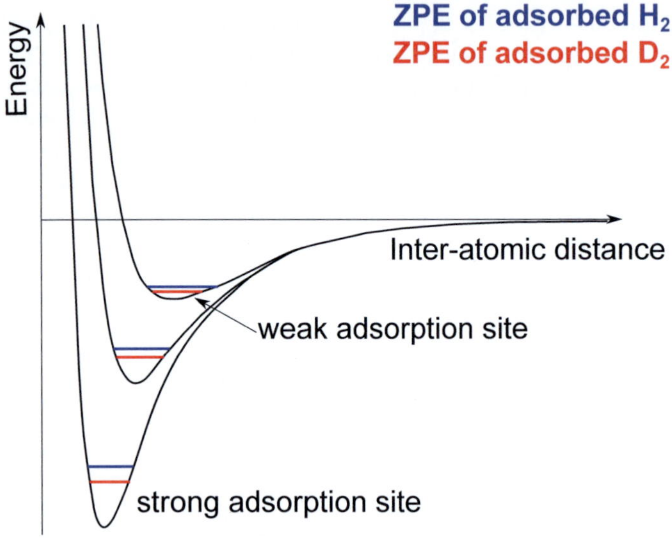

Figure 2.5: The different depths of the potential well represent the strength of the binding site. The zero-point energy of adsorbed hydrogen isotopes differ most on strong sites.

Figure 2.5 shows the impact of different adsorption strengths on the zero-point energy of hydrogen and deuterium. For a strong binding site the difference between the zero-point energies of the hydrogen isotopes is higher in comparison to a weak adsorption site, which results in a higher adsorption energy for the heavier isotope. Deep adsorption potentials lead to a high desorption temperature in the thermal desorption spectroscopy spectrum and the curvature of their potential well affects the selectivity. [44]

2.3 Thermal desorption spectroscopy

The thermal desorption spectroscopy (TDS) technique is a tool widely used in surface science and for the investigation of catalytic materials. [45–47] The experimental setup and procedure is described in chapter 3.1. For the TDS experiment a temperature program with a linear heating rate is applied to the material (adsorbent), which has been previously exposed to a gas at a given exposure temperature. During the temperature increase, the partial pressure of different gases is detected by a mass spectrometer in a continuously pumped system. These partial pressures are plotted versus the measured temperatures, resulting in a TDS spectrum. Considering a pumping rate high enough to prevent re-adsorption, each partial pressure is proportional to the desorption rate of the corresponding gas and the area under the full spectrum is proportional to the total amount of gas, which has been initially adsorbed at the exposure temperature. A TDS spectrum consisting of multiple desorption maxima implies that the material has several adsorption sites with distinct binding strengths. Respecting the principle of microscopic reversibility the adsorption and desorption process can be specified by the same set of rate equations. For one desorption maximum, the desorption rate r_{des} is described by a rate equation of nth order:

$$r_{des} = -\frac{d\Theta}{dt} = k_n \cdot \Theta^n \tag{2.5}$$

where Θ is the instantaneous coverage and k_n is the rate constant, which gives the probability that a physisorbed gas molecule could escape back to the gas phase. The rate constant can be described by the *Arrhenius* equation:

$$k_n = \nu_n \cdot exp\left(\frac{-E_{des}}{RT}\right) \tag{2.6}$$

with ν_n as frequency factor, E_{des} is the desorption energy, R the gas constant and T the temperature. Combining eq. 2.6 and eq. 2.5 leads to the *Polanyi-Wigner* equation [48]:

$$r_{des} = -\frac{d\Theta}{dt} = \nu_n \cdot exp\left(\frac{-E_{des}}{RT}\right) \cdot \Theta^n \tag{2.7}$$

Combining this equation with the linear heating rate, β, and an initial temperature, T_0,:

$$T = T_0 + \beta t \tag{2.8}$$

yields:

$$-\frac{d\Theta}{dT} = \frac{\nu_n}{\beta} \cdot exp\left(\frac{-E_{des}}{RT}\right) \cdot \Theta^n \tag{2.9}$$

Figure 2.6 depicts the combination of the fractional coverage $\Theta(T)$ and the rate constant $k_n(T)$ to the TDS spectrum.

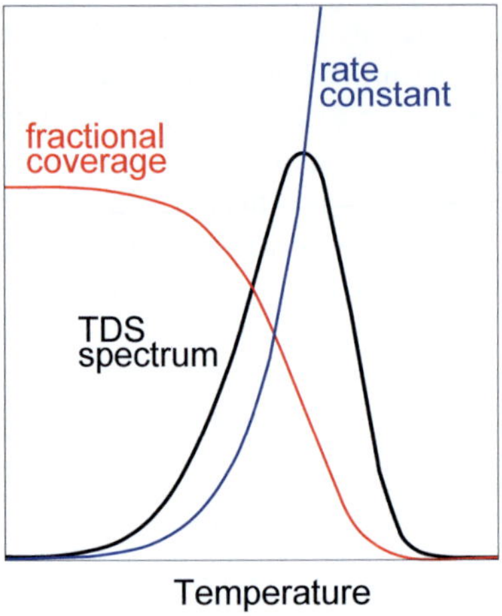

Figure 2.6: The measured desorption rate (TDS spectrum) is composed of the temperature dependent coverage and the rate constant.

Measuring the desorption for different initial coverages allows to determine the reaction order by the concentration dependence of the maximum temperature. Figure 2.7 depicts TDS spectra of different desorption orders n with different initial coverages.

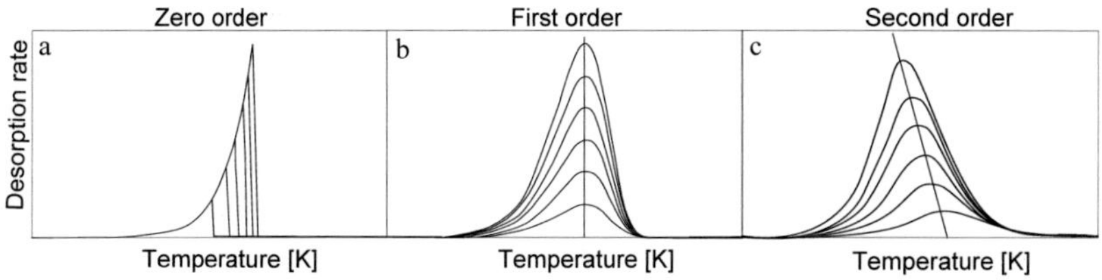

Figure 2.7: TDS spectra of different desorption orders. For higher coverages, the maximum temperature of the desorption rate shifts to (a) higher temperatures for zero-order desorption, (b) has no shift for first-order desorption and (c) shifts to lower temperatures for second-order desorption.

- Zero-order desorption process ($n = 0$) is described by the equation:

$$-\frac{d\Theta}{dT} = \frac{\nu}{\beta} \cdot exp\left(\frac{-E_{des}}{RT}\right) \tag{2.10}$$

In this equation the desorption rate does not depend on coverage ($\Theta(T)$) and it increases exponentially with temperature until a rapid drop occurs when all molecules have been desorbed. The temperature of the peak maximum moves to higher temperatures with higher initial coverages (Θ_0). Zero-order kinetics is common if the adsorbate forms multilayers.

- First-order desorption process ($n = 1$) is described by the equation:

$$-\frac{d\Theta}{dT} = \frac{\nu}{\beta} \cdot exp\left(\frac{-E_{des}}{RT}\right) \cdot \Theta \tag{2.11}$$

In case of first-order desorption the desorption rate is proportional to the fractional coverage ($\Theta(T)$). The temperature of the maximum is independent of the initial coverages. This desorption order represents the case of non-dissociative molecular and atomic adsorption.

- Second-order desorption process ($n = 2$) is described by the equation:

$$-\frac{d\Theta}{dT} = \frac{\nu}{\beta} \cdot exp\left(\frac{-E_{des}}{RT}\right) \cdot \Theta^2 \tag{2.12}$$

The desorption rate of the second-order process is proportional to the square of the fractional coverage. The maximum temperature of the peak moves to lower temperature with increasing initial coverage. This desorption order is observed if re-combinative desorption occurs.

Extracting the desorption energy from TDS measurements is among other techniques possible with the Kissinger method, initially suggested for the analysis of differential thermal analysis data. [49,50] For this method thermal desorption spectra recorded with different heating rates are required. This technique is based on the assumption that the fractional surface coverage at the maximum is independent of the heating rate, which is a reasonable assumption for most desorption processes. [51,52] The following derivation describes this analysis.

At $T = T_{max}$ the condition:

$$\left.\frac{dr_{des}}{dT}\right|_{T_{max}} = 0 \tag{2.13}$$

must be fulfilled. With

$$r_{des} = -\frac{d\Theta}{dt} = -\beta \cdot \frac{d\Theta}{dT} \tag{2.14}$$

equation 2.13 becomes:

$$0 = \left.\frac{d^2\Theta}{dT^2}\right|_{T_{max}} = n \cdot \Theta^{n-1} \cdot \frac{d\Theta}{dT} + \Theta^n \cdot \frac{E_{des}}{RT_{max}^2}. \tag{2.15}$$

Replacing $d\Theta/dT$ with equation 2.9 and rearranging gives:

$$\frac{E_{des}}{RT_{max}^2} = \frac{1}{\beta} \cdot \nu_n \cdot n \cdot \Theta^{n-1} \cdot exp\left(\frac{-E_{des}}{RT_{max}}\right) \tag{2.16}$$

Specifying equation 2.16 with the desorption order leads for $n = 1$ to:

$$\frac{E_{des}}{RT_{max}^2} = \frac{1}{\beta} \cdot \nu_1 \cdot exp\left(\frac{-E_{des}}{RT_{max}}\right) \tag{2.17}$$

and for $n = 2$ to:

$$\frac{E_{des}}{RT_{max}^2} = \frac{2}{\beta} \cdot \nu_2 \cdot \Theta(T_{max}) \cdot exp\left(\frac{-E_{des}}{RT_{max}}\right) \tag{2.18}$$

Rearranging equation 2.17 and taking the logarithm yields:

$$ln\frac{\beta}{T_{max}^2} = -\frac{E_{des}}{RT_{max}} + ln\frac{\nu_1 R}{E_{des}} \tag{2.19}$$

In case of a second-order desorption process the desorption maxima are symmetric with respect to T_{max}, thus $\Theta(T_{max}) = \Theta_0/2$ with Θ_0 as initial coverage. This leads to:

$$ln\frac{\beta}{T_{max}^2} = -\frac{E_{des}}{RT_{max}} + ln\frac{\nu_2 \cdot R \cdot \Theta_0}{E_{des}} \tag{2.20}$$

The reaction order does not have to be determined for the calculation of the desorption energy. The only difference between first and second order is the value of the intercept with the ordinate. For the first order, the frequency factor ν_1 can be determined directly from the intercept, whereas for the identification of ν_2, Θ_0 must be known.

An alternative approach has been proposed by Falconer and Madix [53] using the logarithm of the desorption rate at the maximum temperature (N_p). For a first order desorption process, equation 2.7 leads to:

$$ln\left(-\frac{d\Theta}{dt}\bigg|_{max}\right) = ln\left(N_p\right) = -\frac{E_{des}}{RT_{max}} + ln(\nu \cdot \Theta) \tag{2.21}$$

Plotting the logarithm of the desorption rate at the maximum temperature versus the inverse maximum temperature yields a straight line with a slope proportional to the value for the desorption energy.
The temperature resolution of a spectrum is increased by reducing the heating rate, while the peak height is reduced. [54] Thus, as the sensitivity is limited by the measurement apparatus, a lower bound exists for the heating rate.

2.4 Molecular spectroscopy

Molecular spectroscopy is one of the most important tools in science for the characterization of materials. When a molecule is exposed to electromagnetic radiation, it absorbs photons of certain wavelengths, while others pass unperturbed. The absorbed photon energy excites the molecule from a low energy state to one with higher energy. These specific wavelengths reveal the corresponding energy value of the transition. From the absorbed wavelengths and the extent of their absorption, the nature of the transitions of the molecules is disclosed, which characterizes the composition and structure of the substance.

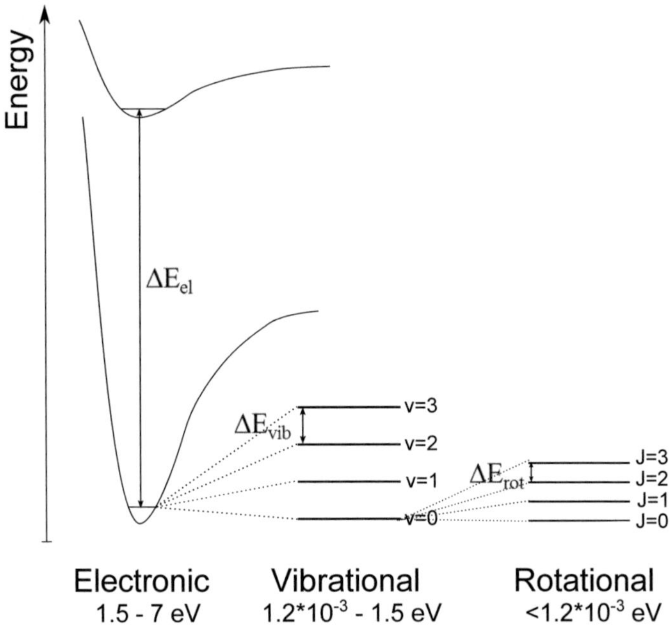

Figure 2.8: Schematic illustration of the molecular transitions. Implemented into electronic states are vibrational energy levels and within these are rotational levels.

The molecule can undergo different transitions depending on the type of electromagnetic radiation it is subjected to. Figure 2.8 depicts the possible transitions. Electronic transitions can occur after irradiation with UV or visible light which can excite an electron to a molecular orbital of higher energy. Mathematically, the vibrational energies of a diatomic molecule can be described by the Morse potential, which is illustrated in figure 2.8. Vibrational transitions are induced by infrared-radiation and rotational transitions occur in the microwave or far infrared region. The following subsections outline the basic principles of Raman spectroscopy and inelastic neutron spectroscopy, which have been used in this thesis for exploring the interaction of molecular hydrogen with MOFs.

2.4.1 Para-ortho hydrogen

Two kinds of molecular hydrogen exist: para-hydrogen with anti-parallel nuclear spin orientation and ortho-hydrogen with parallel nuclear spin orientation. The occupation of the rotational energy levels is affected by these two possible spin configurations. Due to symmetry constrains [55], para-hydrogen occupies rotational levels with even rotational quantum numbers and ortho-hydrogen occupies rotational levels with odd rotational quantum numbers. Therefore, the rotational ground state $J = 0$ of hydrogen can only be occupied by para-hydrogen molecules, while $J = 1$ is the lowest possible state for ortho-hydrogen. In naturally occurring hydrogen, referred to as "normal hydrogen", the ortho to para ratio is 3:1. For normal hydrogen, this ratio is conserved even at cryogenic temperatures, where only $J = 0$ and $J = 1$ are populated. [56] Transitions between para- and ortho-hydrogen occur with a low probability, thus, liquid molecular hydrogen kept below its boiling point converts to para-hydrogen over a period of several days. [34] This process can be accelerated when the liquid is brought in contact with a paramagnetic catalyst leading to conversion time constants of seconds to minutes. Both, para- and ortho-hydrogen can be prepared with purity higher than 99%. [57]

2.4.2 Inelastic neutron spectroscopy

Inelastic neutron spectroscopy is a technique similar to Raman spectroscopy. The scattered particles here are neutrons, whereas they are photons in the case of Raman spectroscopy. The uncharged neutrons are ideal to probe the structural and dynamical nature of materials. Charged particles are attracted or repelled by electrons of the investigated material before they can penetrate deeply. In contrast, neutrons pass easily through the metal walls of sample holders (like aluminum or stainless steel), which makes them suitable to measure bulk properties. They interact with atomic nuclei via the short-range

strong nuclear force. Thus, solid matter is not very dense from a viewpoint of a neutron. The scattered neutron transfers momentum to the atom and the rate of scattered neutrons is proportional to the concentration of elements in the sample and their respective scattering cross section.
When the scattered neutron waves from different nuclei interfere, the position of the atoms in a structure can be determined. Coherent elastic neutron scattering is used in diffraction experiments, one example is the localization of deuterium in crystalline materials. [58] The scattering is called incoherent, when no interference happens. The reason for this is a natural isotopic and spin mixture of the sample, i.e. when no local nuclear spin order exists. This is true, when different nuclear spin isotopes or different spin orientations are present in the system. [34] An energy exchange between the investigated system and the neutrons leads to inelastic scattering. In this work, the adsorption of hydrogen has been studied with inelastic neutron scattering (INS). INS is especially suitable to study hydrogen due to its uniquely high incoherent scattering cross section of 80 barn. [59] The hydrogen visibility for INS is approximately ten times higher than that of any other element (≤ 5 barn). [60] The rotational transitions of molecular hydrogen lie in the low energy transfer region and are observable with INS around the elastic line occurring from elastic scattered neutrons. The rotational energy levels of an isolated hydrogen molecule can be described by [55]:

$$E_{J,M} = B_{rot} \cdot J \cdot (J+1) \tag{2.22}$$

where J is the rotational angular momentum quantum number and $M = 0, \pm 1, .. \pm J$ the magnetic quantum number. B_{rot} is the rotational constant which is inversely proportional to the moment of inertia. Hydrogen has the lowest possible momentum of inertia, thus, the highest rotational constant of $B_{rot} = 7.35\,\text{meV}$. [34]

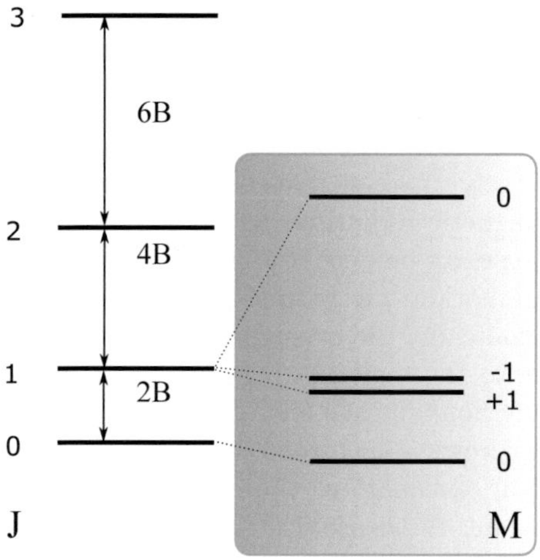

Figure 2.9: Schematic illustration of the rotational transitions and the quantized nature of the z-components of the angular rotational momentum vector (M-levels).

The rotational states are degenerate if the molecule is not affected by a potential field (figure 2.9 left). The degeneracy can be lifted (figure 2.9 right) for example if the molecule is adsorbed on a surface. The quantum number M represents the quantized nature of the z-components of the angular momentum vector J, conventionally aligned along the laboratory axis z.

2.4.3 Raman spectroscopy

The Raman effect describes, how photons are inelastically scattered by a sample. It has been discovered in 1928 by C.V. Raman to appear in liquids and by G. Landsberg and L. Mandelstam to appear in solids. Since then, scientists found an ever-increasing number of applications including, as a tool for the detection of phonons, plasmons, magnons and superconducting gap exitations, and for the observation of transitions in molecules.
When a sample is irradiated by monochromatic light of typically visible or near-infrared frequencies ν_0, the majority of the scattered light will also be of frequency ν_0 and hence can not be discriminated from the incident, high power laser beam. This so called Rayleigh line is the result of elastically scattered photons. However, a very small amount of light is scattered inelastically, and

appears in the spectrum at specific frequencies above and below the Rayleigh line. The frequencies, intensities and polarizations of this light can give valuable information about many different sample properties.
Major properties of Raman scattering can be explained from a classical viewpoint. The positively charged nuclei and the electrons get distorted by the incident electromagnetic radiation, which separates the charge centers of the molecule. This causes an induced dipole moment and the molecule is polarized. When the internal motion (i.e. vibration, rotation) of the molecule changes the polarizability periodically, the frequency of the scattered light is changed by a frequency specific to the molecule ($\nu_0 \pm \nu_{vib/rot}$). This effect even occurs for non-polar diatomic homonuclear molecules like H_2, which cannot be observed with infrared or microwave spectroscopy due to the absence of a dipol moment. Beside the classical explanation, the occurrence of Raman scattering can be described by a quantum theoretical approach. As illustrated in figure 2.10, monochromatic photons, which undergo collisions with the molecule, lift the molecule into a virtual state of higher energy. After this excitation, there are three possibilities for the molecule:

I decay in a higher vibrational state and energy gain from the photon (inelastic scattering: Stokes),

II decay in the initial state (elastic scattering: Rayleigh),

III decay in a lower vibrational state and the photon gains energy from the molecule (inelastic scattering: anti-Stokes)

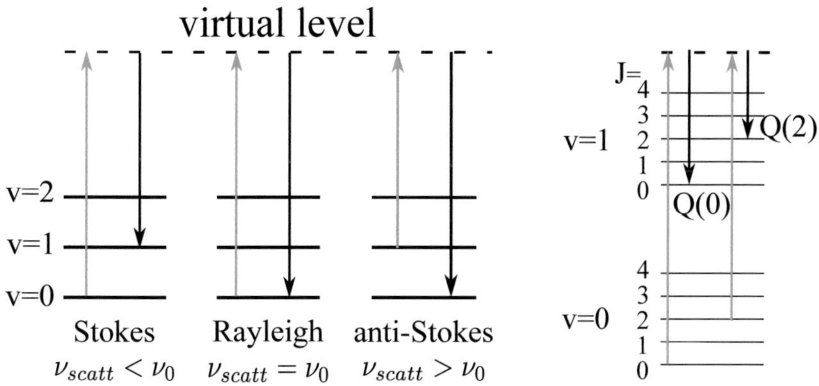

Figure 2.10: Left: schematic illustration of the vibrational Raman scattering and Rayleigh scattering. The molecule gets excited in a virtual level and decays in the same level (Rayleigh), in a higher level (Stokes) or in a lower level (anti-Stokes). Right: Stokes transitions, with ΔJ= 0 (Q(J)-lines). The final and initial rotational quantum numbers J are identical and depicted in brackets.

The photon energy changes by the energy difference between initial and final state, which is shown in eq. 2.23:

$$h\nu_0 - h\nu_{scatt} = \Delta E_m \tag{2.23}$$

Here h is Planck's constant, ν_0 the frequency of the incident light, ν_{scatt} the frequency of the scattered light and ΔE_m the energy change of the molecule owing to vibrational or rotational transitions. The frequency of the incident light is set as zero wavenumber reference ($\nu_0 = 0\,\text{cm}^{-1}$) and the change of the photon frequency is directly detected. Therefore Raman spectra are denoted with "Raman shift" in wavenumbers (cm^{-1}), whereas $1\,\text{cm}^{-1}$ is approximately $0.124\,\text{meV}$.

During a vibrational transition, the change of the rotational quantum number J is not excluded. In this thesis, Raman spectroscopy has been used to investigate the Q(J)-branches of hydrogen and deuterium molecules. These transitions are pure vibrational transitions ($\Delta J = 0$), thus rotational transitions are not further discussed in this context. On the right side of figure 2.10 the Q(J) transitions are shown. The rotational quantum number, which is written in brackets, depicts the initial and final rotational state.

3 Experimental methods

3.1 Thermal desorption spectroscopy

3.1.1 TDS setup

This chapter gives a detailed description of the TDS apparatus developed at the Max Planck Institute for Intelligent Systems, Stuttgart. In order to obtain a low temperature TDS spectrum, three main components are required:

1. a system to deliver gas to the sample
2. a system to cool or heat the sample
3. a mass spectrometer to analyze the amount and type of the desorbed molecules

Figure 3.1 illustrates the main components of the apparatus. The sample holder (red) is mounted into the sample chamber (yellow) from the top with a removable rod. The sample holder is made of Cu and is screwed tightly to the Cu block at the bottom of the sample chamber, which consists of Cu and a CuBe alloy. The temperature of the sample is regulated by a electric heating spiral and a liquid helium flow cryostat, which are connected via a Cu heat exchanger. This connection can be decoupled if temperatures above 300 K are required, e.g., during the activation process. The temperature is detected at two positions each equipped with a pair of a CrNi thermocouple and a Pt resistance sensor. One pair is placed at the bottom of the cryostat and the other is directly glued on the Cu block of the sample chamber. The operating temperature range lies between 20 K and approximately 600 K. The cryostat and the sample chamber are surrounded by an insulating vacuum and a liquid nitrogen shield.

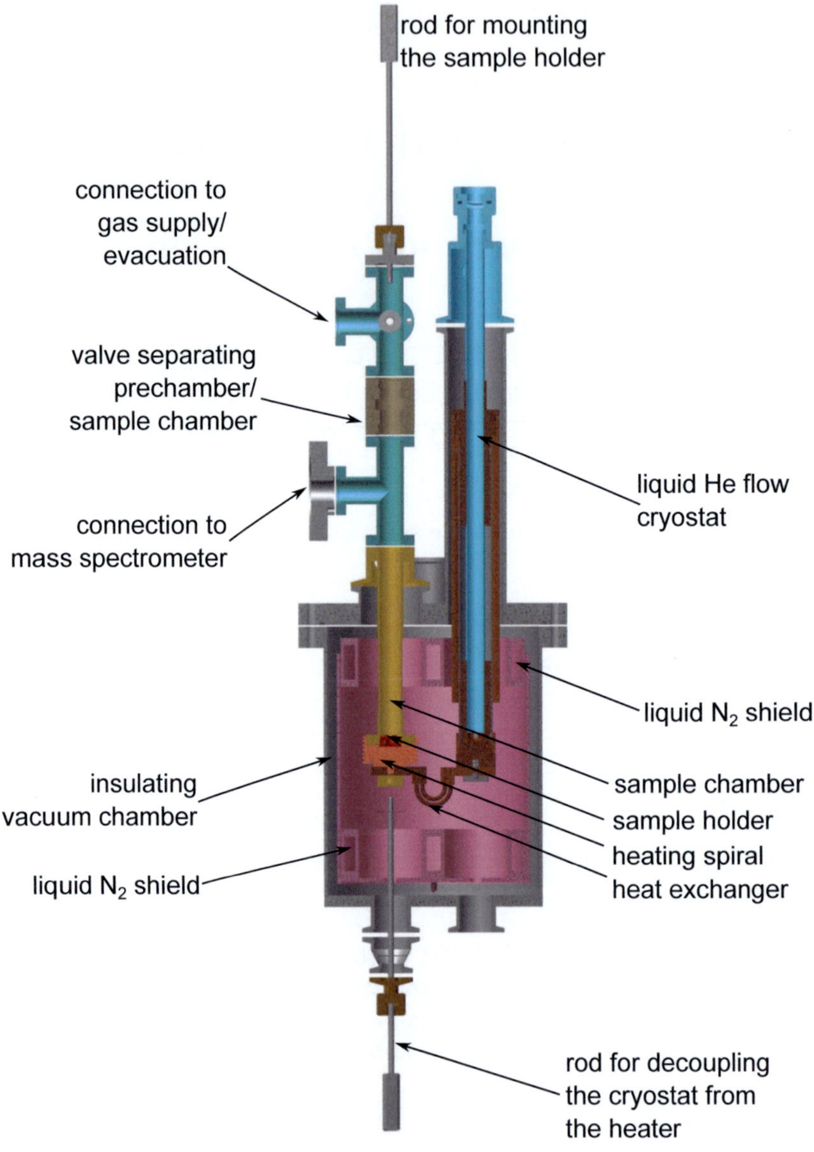

Figure 3.1: Technical drawing of the thermal desorption spectrosocopy apparatus. [61]

3.1.2 Calibration of the TDS apparatus for hydrogen isotopes

The mass spectrometer measures partial pressures of the desorbing gas species. In order to have a quantitative measurement of the desorbed hydrogen isotopes, the signal of the mass spectrometer has to be calibrated. A $Pd_{95}Ce_5$ alloy loaded with a known amount of hydrogen or deuterium is desorbed in the TDS apparatus. The diatomic molecules dissociate on the Pd surface before getting absorbed. The Ce atoms act as trapping centers, which offer the possibility to handle the sample for several minutes under ambient conditions for determining the hydrogen content by weighing. [62–65] On the Pd surface the atoms recombine to diatomic molecules and desorb.

Prior to the calibration process, approximately 0.5 g of $Pd_{95}Ce_5$ have been etched by aqua regia in order to remove the oxide layer on the surface. For removing hydrogen atoms, which have been absorbed during the etching procedure, the sample is heated up from room temperature to 600 K with a heating rate of $0.1\,\mathrm{K\,s^{-1}}$ under high vacuum in the TDS apparatus. The calibration sample has been dismounted after it reached room temperature for weighing and then mounted back into the sample chamber of the TDS apparatus. By exposing the $Pd_{95}Ce_5$ alloy to H_2 or D_2 atmosphere of 150 mbar at 350 K for 2 h the sample is loaded. After cooling to room temperature, the sample is weighted outside the sample chamber to quantify the amount of absorbed gas. This weighing procedure is performed quickly and the alloy is mounted back into the sample chamber. Then, the temperature controlled desorption procedure with a heating rate of $0.1\,\mathrm{K\,s^{-1}}$ from room temperature to approximately 600 K is started. An example of the detected signal for two different calibration measurements is depicted in figure 3.2.

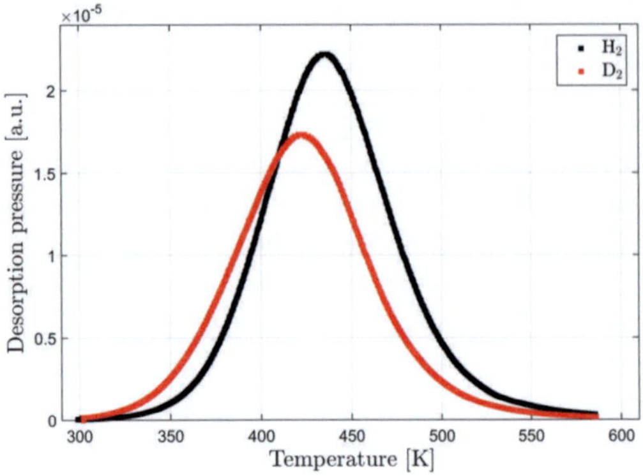

Figure 3.2: Desorption spectra of the calibration procedure. The $Pd_{95}Ce_5$ alloy is heated to 350 K and then exposed to 150 mbar of hydrogen or deuterium for 2 h.

All absorbed hydrogen or deuterium desorbes with a maximum at about 435 K and 425 K, respectively. The area under the desorption maximum is directly related to the amount of hydrogen or deuterium previously absorbed by the alloy, which has been determined by the weight difference between prior and after loading. This calibration has been performed several times to obtain a reliable calibration constant for the output of the mass spectrometer in molecules per second. Any change of the mass spectrometer settings, repair work or exchange of the filament require new determination of the calibration constants.

3.1.3 Experimental procedure for TDS

Every measurement series starts with the activation of the sample with a typical mass of 2 mg-8 mg, which means heating the adsorbent under high vacuum conditions in the TDS device. This process is necessary to assure that physisorbed molecules like impurities adsorbed from air during the loading procedure or remains from the synthesis are removed. The applied activation temperature depends on the thermal stability of each sample. The activation time is chosen by monitoring the pressure of the outgassed molecules with the mass spectrometer. In this work, activation temperatures between 373 K - 500 K and activation times between 1 h - 6 h have been applied.
To get an overview of the adsorption behavior of a new sample, the first measurements are performed with pure gas, loaded at room temperature and immediately cooled down to the lowest temperature that can be reached under current experimental conditions. This temperature is called base temperature and is usually around 20 K. At this temperature the non-adsorbed gas is evacuated and the temperature ramp with $0.1\,\mathrm{K\,s^{-1}}$ starts. The desorption of the molecules is monitored by the mass spectrometer (see figure 3.3a). This loading procedure gives a first clue on how the exposure parameters have to be chosen for the following measurement series of the given material. The exposure temperature in this work has been varied between 20 K and 120 K, the exposure times between 5 min and 180 min and the exposure pressures depend if mixture loading (maximum 60 mbar) or pure gas loading have been applied (maximum 900 mbar).
Depending on the material, the evacuation of the non-adsorbed gas has been performed at exposure temperature (see figure 3.3b) or base temperature (see figure 3.3c).

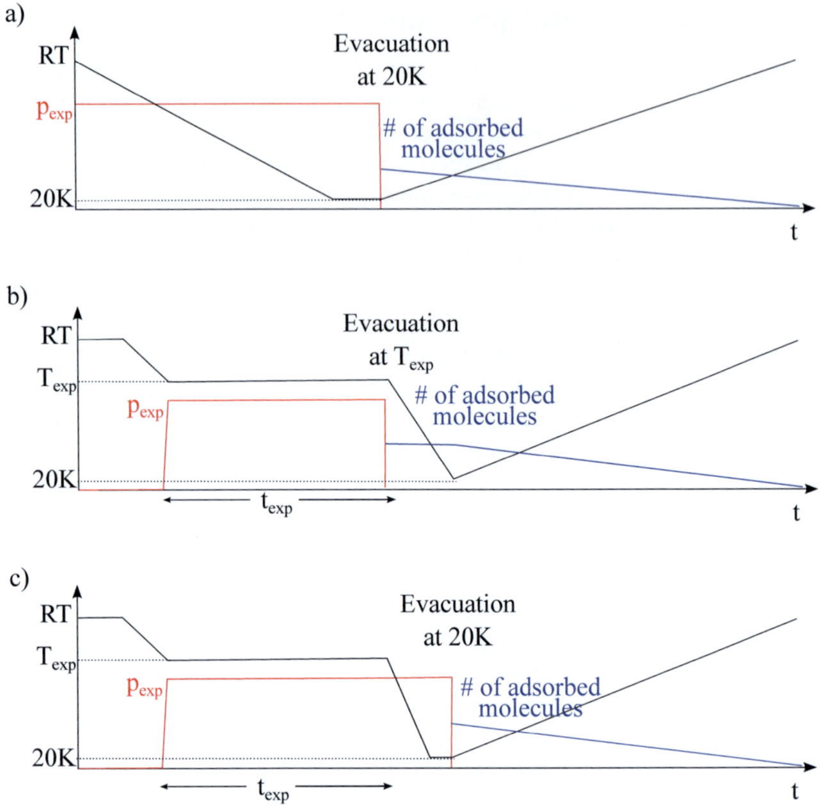

Figure 3.3: Temperature (black) and pressure (red) profiles for different measurement procedures. After starting the TDS measurement, the pressure profile is not shown anymore, instead the number of adsorbed molecules is depicted in blue. a) Gas-loading at room temperature and evacuation of the non-adsorbed gas at 20 K (*base temperature*). b) Gas-loading for an *exposure time* after the *exposure temperature* has been reached and evacuation of the non-adsorbed gas at *exposure temperature*. Cooling to *base temperature*. c) Gas-loading for an *exposure time* after the desired *exposure temperature* has been reached and evacuation of the non-adsorbed gas at *base temperature*. The temperature ramp always starts from *base temperature*. Please note that the pressure drop with decreasing temperature has not been taken into account in this schematic diagram.

Isotope mixture and selectivity

The equimolar hydrogen isotope mixture has been produced outside the TDS chambers. Two identical volumes containing the same pressure of hydrogen and deuterium have been prepared and afterwards simultaneously connected to the gas line, which is connected to the TDS prechamber. This procedure has been done before each measurement. The desorption experiment is performed identical to the pure gas experiment.
Due to the mass spectrometer the unique possibility to determine simultaneously the desorbing isotopes is given. Mass three has also been detected to ensure if dissociation took place, which leads to forming of HD molecules. After calibration of the mass spectrometer signals, the selectivity can be directly determined and is given for a mixture of component A and B by [66]:

$$S(A,B) = \frac{x(A)/x(B)}{y(A)/y(B)} \tag{3.1}$$

where x (y) refers to the mole fraction of particles in the adsorbed (gas) phase. The 1:1 ratio of the gas mixture remains nearly unchanged during the adsorption process, since the adsorbed gas volume is very small compared to the volume of the sample chamber ($y(A)/y(B) \approx 1$). Therefore, the selectivity is determined by the ratio between the adsorbed amount of isotope A and isotope B.

Stepwise-loading procedure

A 'stepwise-loading procedure' of the isotopes was developed to investigate possible isotope exchange. Figure 3.4 shows a schematic illustration. Isotope 1 is introduced to the combined volume of prechamber and sample chamber with a fixed exposure pressure. The valve connecting the pre- and sample-chamber is closed and the gas in the prechamber gets evacuated. After isotope 1 has been exposed to the sample for a certain exposure time, isotope 2 is introduced to the sample. For loading isotope 2 both chambers have to be connected again. In this step, the non-adsorbed amount of isotope 1 in the sample chamber expands to the prechamber. Isotope 2 is added with the initial exposure pressure of isotope 1. Again, the connection between the chambers is closed and the prechamber gets evacuated. The mixture in the sample chamber has the ratio of 1:2.5 between (isotope 1):(isotope 2). At the end of the exposure time the non-adsorbed isotopes are evacuated, the sample is cooled to base temperature and the desorption experiment starts.

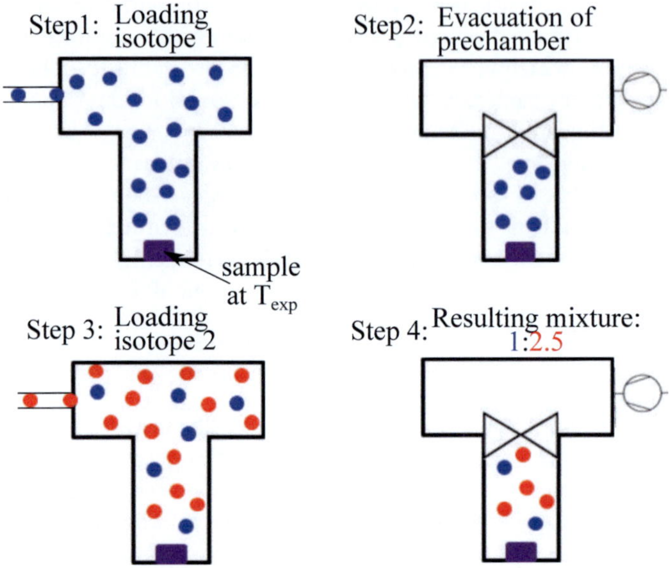

Figure 3.4: Visualization of the stepwise-loading procedure.

3.2 Inelastic neutron scattering

The INS experiments have been performed at the Spallation Neutron Source (SNS) of the Oak Ridge National Laboratory on beam line 16B [67]. For neutron production by spallation an accelerator-based system provides short proton pulses in the range of microseconds, which strike a heavy-metal target (liquid mercury) at almost the speed of light in vacuum. Only a few neutrons are produced directly by the impact, the more important process is the excitation of heavy-metal atoms, which then evaporate 20-30 neutrons per nucleus. A moderator (liquid H_2, D_2O) slows the neutrons down before they are guided to the beam lines. The indirect geometry spectrometer VISION, allocated at beam line 16B, uses a wide range of incident neutron energies. The energy resolution of $\Delta E/E \approx 1 - 2\%$ is constant to good approximation over the entire energy transfer range.
A sample mass of 1.044 g of activated Cu(I)-MFU-4*l* was loaded under helium-atmosphere into an aluminum can, which was mounted afterwards on a gas-handling system. The sample has been evacuated by a turbo-molecular pump at room temperature overnight. Inelastic neutron spectra of the blank sample and gas-loaded sample have been recorded at temperatures from 5 K to 200 K. The temperature has been adjusted through heaters attached to the sample cell and a closed cycle helium refrigerator. Prior to the volumetric loading procedure of H_2 at 70 K, it has been converted to para-H_2 at 20 K by using Fe_3O_4 as catalyst.

3.3 Raman spectroscopy

The experimental Raman setup employed for measurements on free and adsorbed hydrogen (deuterium) is shown schematically in figure 3.5a. For the excitation of the molecule the 514.5 nm line of an Argon-ion laser was used. The weaker plasma lines of the laser were spectrally filtered with a grating pre-monochromator. In order to avoid heating effects which lead to possible sample degradation, the incident laser power was always kept below 5 mW. The incident laser beam was guided with mirrors to the sample where it has been focused through a photographic objective to a $\sim 50\,\mu$m spot on the sample surface. The same objective was also used for the collection of the scattered light. For the suppression of stray light not coming from the sample the scattered light had to pass through a pinhole (positioned at the image plane of the

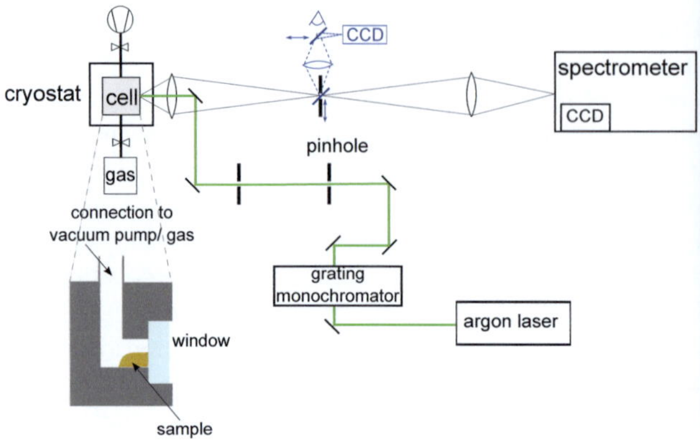

(a)

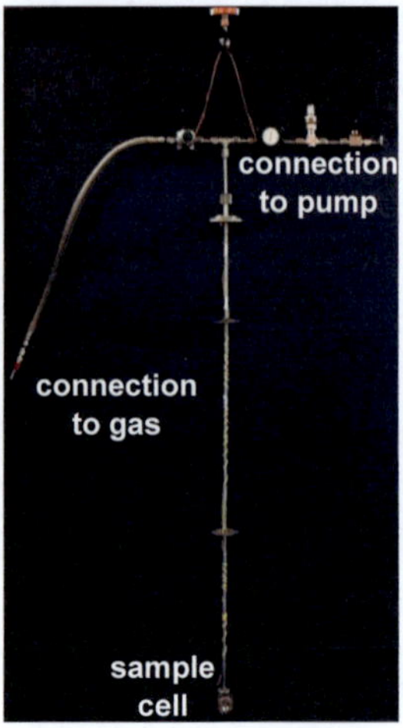

(b)

Figure 3.5: Schematic illustration of the Raman experimental setup and picture of the sample holder.

sample). For adjusting the position of the incident laser beam on the sample, the sample can be directly observed by additional mirrors and an objective lens. The scattered light is focused by a lens on the entrance slit of a Jobin-Yvon triple/ or a single grating spectrometer, equipped with a nitrogen-cooled CCD camera (Jobin-Yvon Symphony). A notch filter was used to suppress the elastically scattered light. The measurements were performed in the single grating configuration yielding a spectral resolution of $1.9\,\mathrm{cm}^{-1}$ in the spectral region of interest. After each measurement a spectrum of an Ar lamp was recorded as reference.

The high pressure experiments have been performed with an especially designed cell made of stainless steel equipped with a suprasil window allowing for optical measurements. [68] The cell fulfills the requirements of low temperatures, high pressures and vacuum. As indicated in figure 3.5a only half of the cell was filled with the sample, enabling to measure the free gas signal (above the sample) and the adsorbed gas signal (in the sample) under the same conditions. The sample was activated at 453 K for 3 h and loaded in the pressure cell under the inert atmosphere of an Ar-glovebox. Afterwards the sample holder had to be mounted to the gas loading (gas evacuation) pipe (see figure 3.5b) outside the glovebox, which causes a short exposure to air. Right after the sample holder was assembled it was evacuated by a turbo molecular pump (over several days) applying high vacuum conditions prior to the gas loading. The measurements have been performed at different temperatures and for each temperature value, the pressure was increased in steps from 1 bar to 22 bar. The gas supply was directly connected to the cell and the applied pressure has been regulated with a pressure reducer (max. pressure 25 bar).

Low temperatures (down to 40 K) have been obtained by immersing the cell in a helium bath cryostat (Konti-IT, Cryovac) equipped with a silicon-diode temperature sensor. The cryostat consists of four chambers: 1. a vacuum chamber isolates the cryostat from room temperature, 2. a pre-cooling chamber filled with liquid nitrogen (down to $\sim$ 80 K), 3. a helium reservoir (isolating vacuum between chamber two and three), 4. the sample chamber, which is connected through a needle valve to the third chamber, allowing controlled helium flow around the sample chamber. The temperature has been stabilized with an electrical heater installed near the sample chamber.

For carefully chosen temperature and pressure combinations, Raman spectra were recorded on the sample and in the free gas above the sample.

4 Metal-organic frameworks

Metal-organic frameworks (MOFs) have emerged over the last two decades, representing a class of nanoporous materials. [69–71] Due to their construction principle, MOFs are also referred to as porous coordination polymers (PCPs). Connecting metal ions by organic ligands was already established in the end of 1980s, beginning of 1990s by the group of Robson. [72–74] Due to the synthesis process these frameworks contain solvent molecules. Unfortunately, the treatment with mild thermal conditions in vacuum for removing the solvent molecules damaged the porosity of the frameworks. The real breakthrough in MOF chemistry has been achieved by O. Yaghi in 1995, who coined the name MOF (first IRMOF-isoreticular MOF). [75,76] His seminal work [77] brought an exceptionally robust framework, which does not collapse even after removal of the solvent molecules. The basic idea was to combine carboxylate rigid organic ligands with metal-ions, which form a strong metal-oxygen bond.

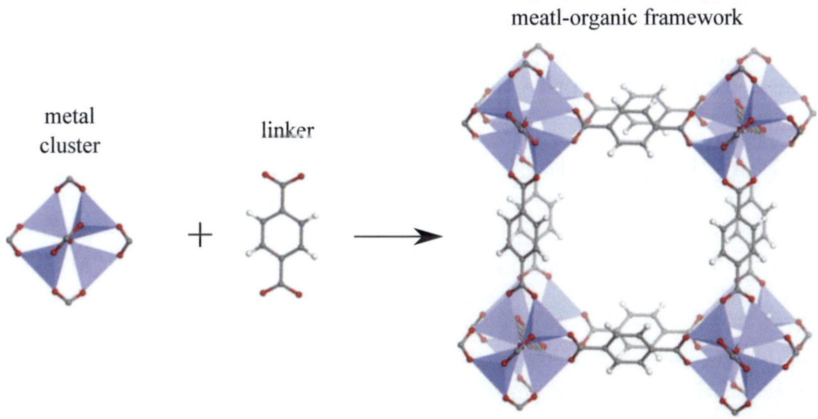

Figure 4.1: Construction principle of MOF-5. The metal cluster with blue polyhedra indicate the ZnO_4 tetrahedra. The linker is a hydrocarbon chain. (Zn: blue, O: red, C: grey, H: white) [78]

The metal-carboxylate subunits build the nodes of a cluster and are connected by organic linkers, which act as spacers to create a regular porous three-dimensional structure. The resulting crystalline framework has low density and high specific surface area. MOF-5 is the best known sample and requires simple and cheap starting materials for the production. In this material, 1,4-benzenedicarboxylate is linked with zinc oxide clusters. [79] Figure 4.1 shows MOF-5 as an example for the general construction principle of a MOF. A common way to synthesize MOFs is to use hydrothermal or solvothermal synthesis, for which the building units are mixed in solution. The self-assembling of metal ions with organic ligands via coordination bonds happens in closed vessels under autogenous pressure above the boiling point of the solution. [80] After the frameworks have grown they precipitate as powder. The framework structure of the crystalline MOFs can be determined with X-ray diffraction. [81] A huge variety of MOFs with different compositions can be synthesized since metal centers and organic linker can be exchanged. Therefore, MOFs are interesting for many different research topics, like inorganic chemistry, material science and pharmacology. They can posses functional properties such as magnetism [82], electric conductivity [83] and luminescence [84–86], which are of great interest for applications in many technologically relevant fields. MOFs have a permanent porosity and large surface areas up to $\approx 7000\,\mathrm{m^2\,g^{-1}}$ [87]. Among crystalline materials, MOFs have the lowest density ranging from 1 to $0.21\,\mathrm{g\,cm^{-3}}$ and consist of up to 91% free volume. [79] Therefore, MOFs become interesting for catalysis, gas separation and for gas storage as well. [88–92] The materials used in this thesis are described in detail in the following.

4.1 IFP-1, IFP-3, IFP-4, IFP-7

Imidazolate **F**ramework **P**otsdam (IFP) is the name of an isoreticular family of microporous MOFs, synthesized by the group of Prof. Holdt from the University of Potsdam. [15, 16] The investigated materials differ by the substituents (Me (IFP-1), Br (IFP-3), Et (IFP-4), OMe (IFP-7)) at the imidazolate-amide-imidate linker. The metal ions for these materials are Zn^{2+}, which form together with the linker a neutral microporous imidazolate metal-organic framework with 1D hexagonal channels with Zn^{2+} located on the edge of the channels (see figure 4.2). The channel walls are formed by the rigid and planar linker. The substituents point into the hexagonal channels and, thus, determine the effective channel diameters. Considering the van der Waals radii of the substituents, the accessible diameters are IFP-1 (4.2 Å) > IFP-3 (3.1 Å) > IFP-7 (2.1 Å) > IFP-4 (1.7 Å). [15,17] Furthermore, the substituent affects the polarisability and functionalisation of the walls. The highly electronegative Br groups induce higher dipolmoments in the linker of IFP-3 compared to the linker in IFP-1, whereas the flexible methoxy and ethyl groups induce conformational freedom into the IFP-7 and IFP-4 frameworks, respectively. Thus, the flexible groups can act as molecular gates for the gas molecules.

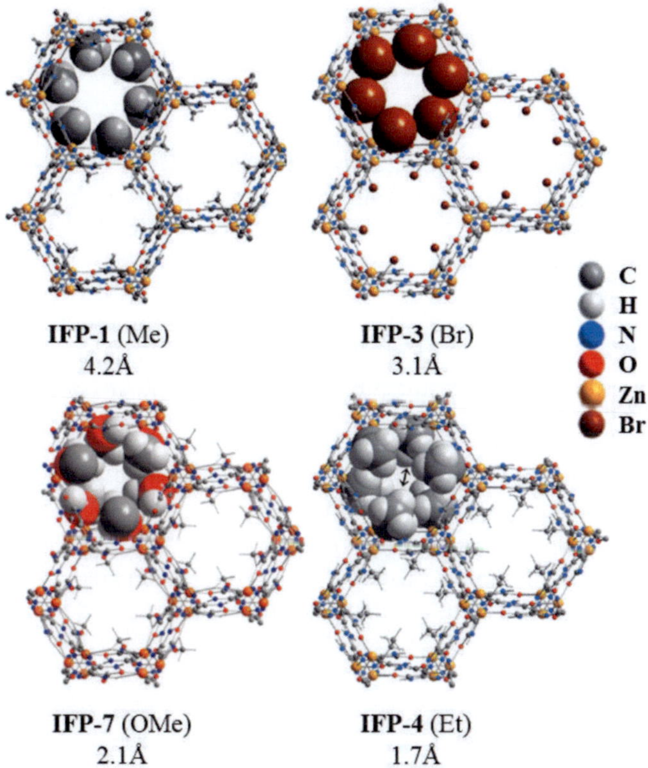

Figure 4.2: The hexagonal channels of the IFP family alter in accessible pore diameter. Different substituents on the linker affect the size, the polarisability and the functionalisation of the samples. [15, 17]

The specific surface area has been calculated with the BET model from adsorption isotherms measured with nitrogen at 77 K. This value decreases in correlation with the decreasing pore diameter IFP-1 ($802\,m^2g^{-1}$) > IFP-3 ($607\,m^2g^{-1}$) > IFP-7 ($15\,m^2g^{-1}$) > IFP-4 ($5\,m^2g^{-1}$). [15,93] From the very low value determined for IFP-7 and IFP-4 it is concluded that nitrogen molecules only populate the outer surface of the material and cannot enter the pores. The geometrical surface area leads to the same descending order: IFP-1 ($728\,m^2g^{-1}$) > IFP-3 ($664\,m^2g^{-1}$) > IFP-4 ($346\,m^2g^{-1}$) and gives a reasonable value for IFP-4. [15]

4.2 MFU-4

The **M**etal-organic **F**ramwork **U**lm University (MFU) has been developed by the group of Prof. Volkmer from Augsburg University, formerly working at the University of Ulm. MFU-4 consists of dianionic $BBTA^{2-}$ linker, which connect the cationic pentanuclear coordination units of $[Zn_5Cl_4]^{6+}$ to a cubic crystal lattice. The synthesis procedure and characterization of the structure is reported by Biswas *et al.* [18]. This series is characterized by alternating small and large pores. Figure 4.3 depicts the atomic structure of Cl-MFU-4. The cavities of the pores are filled with yellow spheres which represent the pore size. The chlorine atoms point into the pore center of the small pore and restrict the diameter to 3.87 Å, whereas pointing out of the large pore leads to a diameter of 11.94 Å. [18] The aperture between the two cavities, formed by 4 Cl atoms, has a diameter of 2.52 Å which excludes large molecules from the framework. Therefore, no nitrogen is adsorbed at 77 K and the specific surface area could not be determined by the standard nitrogen adsorption method. [18] The accessible surface area has been calculated as proposed by Düren et al. [94] to $1611\,m^2g^{-1}$.

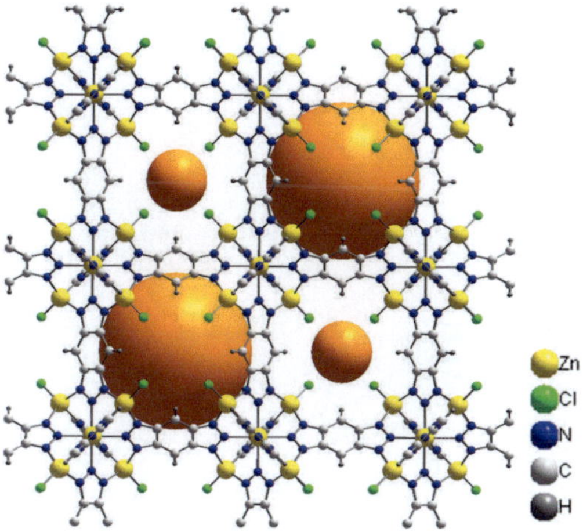

Figure 4.3: Structure of Cl-MFU-4. Chlorine atoms (green) pointing into the pore reduce the volume, resulting in a diameter of 3.87 Å. In consequence, for the neighboring pore the chlorine atoms point out of the pore, resulting in a diameter of 11.94 Å.

The aperture between the two cavities can be altered by replacing the chlorine atoms, which leads to derivatives of the pristine material Cl-MFU-4. Using larger atoms like bromine results in a diameter of 2.33 Å for the pore aperture of Br-MFU-4.

4.3 MFU-4*l* and Cu(I)-MFU-4*l*

MFU-4 large (MFU-4*l*) belongs to the group of isoreticular MFU-4-type frameworks, which consists of the cationic pentanuclear coordination units $[Zn_5Cl_4]^{6+}$. For this compound the cationic units are connected by the $BTDD^{2-}$ linker, which is approximately two times longer then the $BBTA^{2-}$ linker of MFU-4. This modification leads to pores with a diameter of 11.97 Å and 18.56 Å. The aperture increased from 2.52 Å (MFU-4) to 9.13 Å in diameter. Figure 4.4 shows the pore structure with spheres representing the pore size. Details of the synthesis are given by Denysenko *et al.* [95].

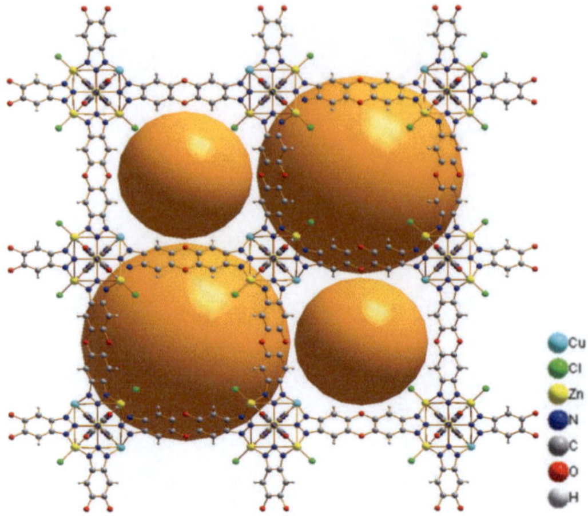

Figure 4.4: Structure of Cu(I)-MFU-4*l*. When the chlorine atoms (green) point into the pore the volume is reduced, resulting in a diameter of 11.97 Å. In consequence, for the neighboring pore the chlorine atoms point out of the pore, resulting in a diameter of 18.56 Å.

The large aperture allows to determine the specific surface area by nitrogen adsorption, to a value of $2750\,m^2g^{-1}$.
Figure 4.4 shows a post-synthetically functionalised MFU-4*l* derivative which contains coordinative unsaturated metal centers of copper. Denysenko *et al.* replaced about 30% of the Zn(II)-Cl units by Cu(I)-ions. [19] This led to threefold-coordinated unsaturated Cu(I) sites with reversible and remarkably strong adsorption of small molecules. For H_2, the heat of adsorption has been determined from adsorption isotherms to be $32.3\,kJ\,mol^{-1}$ and DFT calculations yield $25.0\,kJ\,mol^{-1}$. [19]

4.4 DUT-9

The metal-organic framework DUT-9 has been developed at **D**resden **U**niversity of **T**echnology (DUT) by the group of Prof. Kaskel. The synthesis and structure properties are described in detail by Gedrich *et al.* [22].

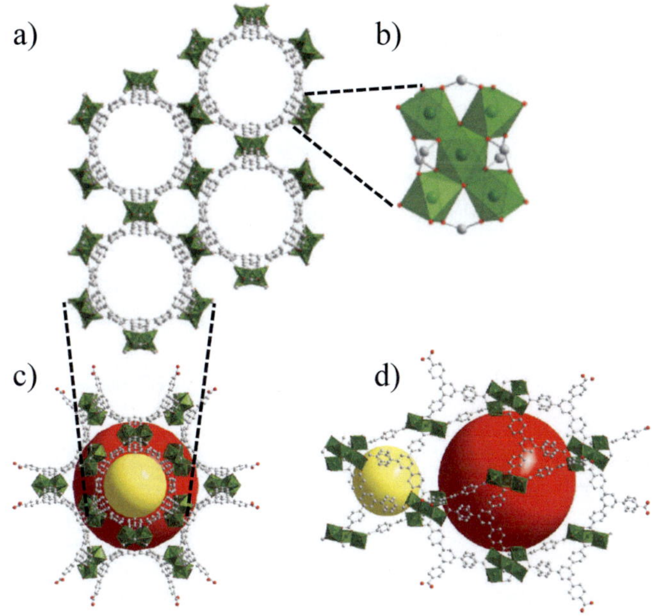

Figure 4.5: a) Plan view of the channels of DUT-9. b) Cluster formed by Ni (green), C (grey), and O (red) atoms. c) Illustration of the pore dimensions in plan view of the channels and d) along the channels (red: 25 Å, yellow: 13 Å in diameter). [22]

This structure is based on $Ni_5O_2(CO_2)_6$ clusters, which are connected by btb (btb = benzene-1,3,5-tribenzoate) linkers forming a 3D network. The cluster, which contains five coplanar Ni atoms (green), is shown in figure 4.5b. The central Ni atom is surrounded by four Ni atoms, which are after activation accessible as open metal sites for the adsorption of guest molecules. This network possesses a bimodal alternating pore structure. The large and the small pore have approximately a diameter of 25 Å and 13 Å, respectively. The pore volumes are illustrated along different crystal axes of the material in figure 4.5c,d as red and yellow spheres. Their arrangement gives rise to hexagonal channels throughout the whole crystal (see figure 4.5a). The surface area could only be determined by the single point BET method to be $3330\,m^2g^{-1}$. [96] Due to the water sensitivity of DUT-9 it should be handled under dry atmosphere.

4.5 CPO-27-Ni

CPO-27, also known as MOF-74, is an isostructural metal-organic framework series, firstly synthesized by Prof. Dietzel in Oslo (**C**oordination **P**olymers of **O**slo). The three-dimensional structure is built of one dimensional honeycomb-like channels. These are filled with water molecules after the synthesis, which can be removed by a mild thermal treatment. [20, 21]

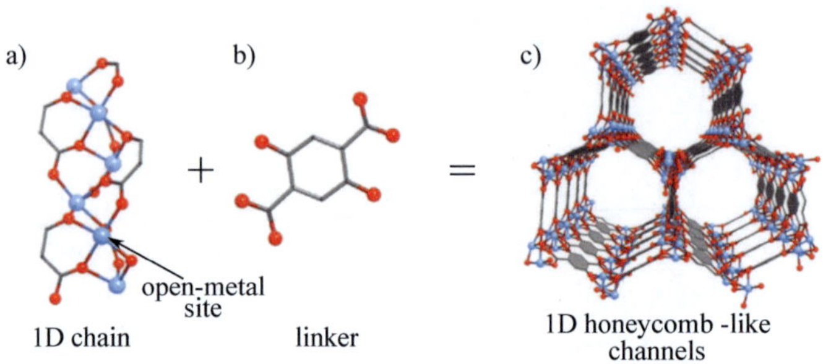

Figure 4.6: Configuration of CPO-27. The pictured chain is already in the water removed condition. The metal center (blue) is coordinated only by five oxygen atoms (red). [97]

The intersections between the honeycomb channels are helical chains of cis-edge connected metal-oxygen coordinations, joined by an organic ligand (figure 4.6b) to form the structure. The metal center, which is located at the corner of the hexagon, is coordinated to six oxygen atoms. One is part of a water molecule pointing towards the cavity (not shown in figure 4.6) and the other five are part of the organic ligand. During the activation process the water molecule attached to the metal center is removed as well, thus the coordination environment of this center changes from octahedral to square pyramidal (figure 4.6a). The helical assembly of the NiO_5 units causes that the nearest nickel atoms are exposed in different channels.
For the investigations of this work $Ni_2(dhtp)(H_2O)_2 \cdot 8\ H_2O$ (dhtp = 2,5-dihydroxyterephthalic acid) with Ni^{2+} as metal component has been chosen. The electron deficient site at the nickel atom is highly reactive and has an isosteric heat of adsorption of 13.5 kJ mol^{-1} for H_2. [98] The channels have a diameter of about 11 Å and the specific surface area of the network is 1083 m^{2}g^{-1} [98].

4.6 Ni-CFA-6

CFA is the abbreviation for **C**oordination **F**ramework **A**ugsburg and is a general term for materials synthesized by the group of Prof. Volkmer from Augsburg University. The CFA-6 series is based on the 4,4'-bipyrazolyl (BPZ) linker, which combines the metal centers M. These materials are labeled M-CFA-6 (also known as M(BPZ)). [23,99] Here, the structure containing nickel as metal center has been investigated. The empirical formula $C_8H_7N_5Ni$ describes the composition of the material. Consecutive metal centers are connected by the linkers, forming 1D linear chains (see figure 4.7). The chains are connected further by the BPZ linkers, which act as spacers. These connections lead to the formation of a 3D framework. The structures crystallize in the orthorhombic crystal system forming linear channels, with a diameter of 5.9 Å. [99]
Figure 4.7 shows the structure without the solvent molecules, which represents the condition after activation. In the as-synthesized state the solvent molecules are only hosted in the channels without being bound to the Ni atoms. [23] The specific surface area is $767\,m^2g^{-1}$ [23].

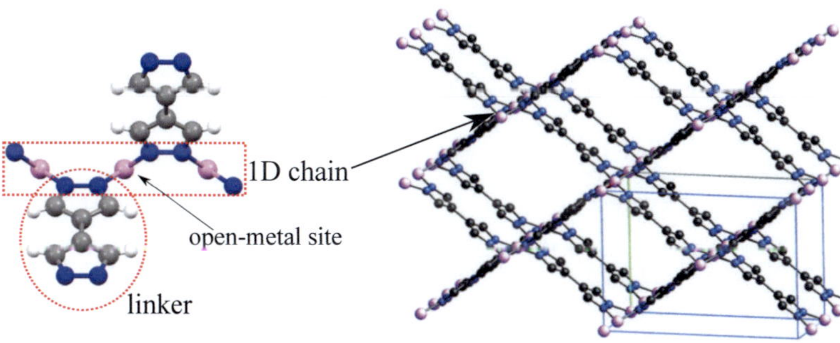

Figure 4.7: Structure of Ni-CFA-6. The metal center nickel is depicted in pink. The BPZ linker consists of nitrogen (blue), carbon (grey) and hydrogen atoms (white). Nitrogen atoms of the linker connect consecutive metal centers and form an 1D chain. H atoms are omitted for clarity in the illustration of the structure on the left side. [23]

5 Isotope adsorption and separation in MOFs with small pores

5.1 Hydrogen isotope adsorption and separation in IFP-1, IFP-3, IFP-4 and IFP-7

5.1.1 Results

In order to study Quantum Sieving in MOFs, a series of IFP samples with different pore sizes has been investigated. Each of the IFP samples has been activated in the sample chamber of the TDS device prior to the first measurement for two hours at 423 K under high vacuum. All desorption spectra have been recorded by applying a heating rate of $0.1\,\mathrm{K\,s^{-1}}$ if not mentioned otherwise.

IFP-1,-3,-4 and -7, have been exposed at room temperature to 10 mbar of either pure H_2 or pure D_2 and after cooling to 20 K under gas atmosphere, the non-adsorbed gas has been evacuated. For each sample the thermal desorption spectrum shows one maximum, whereas the maximum temperature varies for the individual samples (see figure 5.1 and table 5.2). Furthermore, the desorption rate, i.e. the amount adsorbed during exposure, differs strongly between the samples (note different scales of y-axes in figure 5.1 and see table 5.2). The desorption rate of pure D_2 always exceeds the one for H_2. For IFP-1 and IFP-3 the maximum is at lower temperatures and ends at 80 K and 100 K, respectively. Whereas for IFP-7 and IFP-4 the maximum appears above 100 K desorption temperature. Additionally, for IFP-4 and IFP-7 a high desorption signal is visible at low temperatures which can be ascribed to the release of gas condensed in inter-particle voids of the powder. This signal occurs as well for IFP-1 and IFP-3 with comparable magnitude, but is not visible in figure 5.1 due to the different scales.

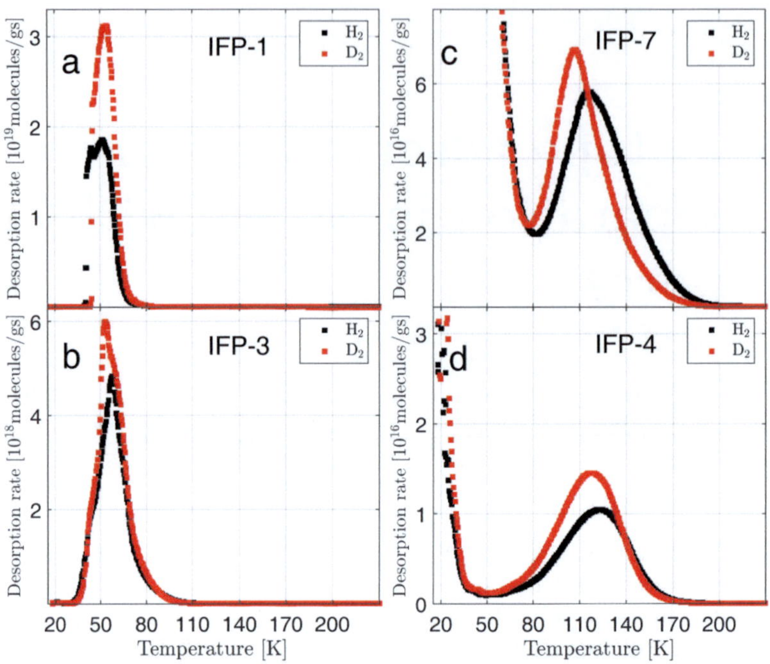

Figure 5.1: 10 mbar of pure H_2 or D_2 have been exposed in the IFP-1, -3, -4 and -7 samples at room temperature. (heating rate: $0.1\,K\,s^{-1}$)

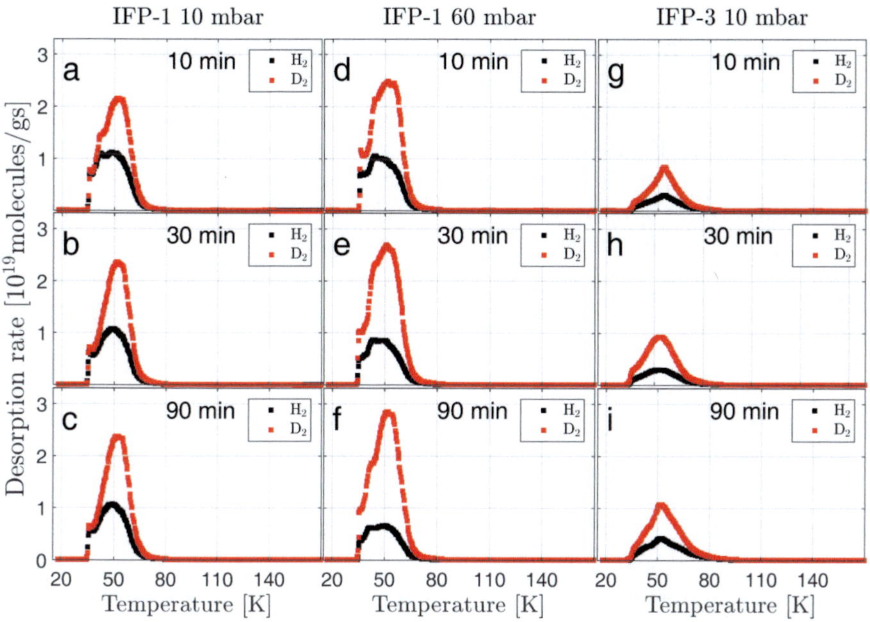

Figure 5.2: An equimolar gas mixture has been applied to IFP-1 and IFP-3 for different exposure times and pressures at 30 K exposure temperature. (heating rate: $0.1\,\mathrm{K\,s^{-1}}$)

The spectra of IFP-1 and IFP-3 measured with pure gas show a similar maximum temperature and a comparable total adsorbed amount (see figure 5.1 and table 5.2). This initiated to study these two materials by applying same conditions. IFP-1 and IFP-3 have been exposed to an equimolar gas mixture at 30 K for different exposure times (10 min, 30 min, 90 min) and the resulting TDS spectra are presented in figure 5.2. The first two columns depict the desorption spectra of IFP-1 exposed to 10 mbar and 60 mbar isotope mixture and the third column shows IFP-3 exposed to 10 mbar isotope mixture. As it was already detected with pure gas, only one desorption maximum appears. For IFP-1 exposed to 10 mbar the ratio between the adsorbed amount of H_2 and D_2 is almost similar for all exposure times. When IFP-1 is exposed to 60 mbar the adsorbed amount of D_2 increases and the adsorbed amount of H_2 decreases for longer exposure times. However, the total amount of adsorbed molecules remains approximately the same for all exposure times. The exposure of IFP-4 and IFP-7 to pure gas revealed a desorption maximum above 100 K. For these materials, longer exposure times (10 min, 60 min, 120 min) have been chosen compared to IFP-1 and IFP-3 due to the diffusion limitation through the smaller pores. Figure 5.3 shows the TDS spectra of IFP-7 exposed to 10 mbar equimolar mixture at 77 K for different exposure times. For IFP-4 different exposure times have been applied at 50 K and 77 K exposure temperature and the spectra are presented in figure 5.4 and 5.5, respectively.

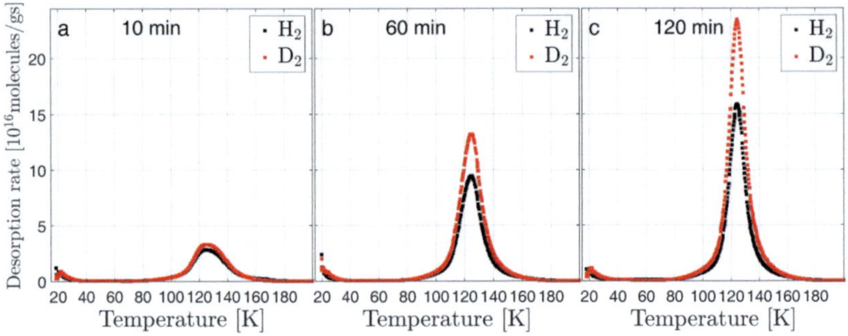

Figure 5.3: TDS spectra of IFP-7 exposed to 10 mbar of an equimolar isotope mixture at 77 K for different exposure times. (heating rate: $0.1\,K\,s^{-1}$)

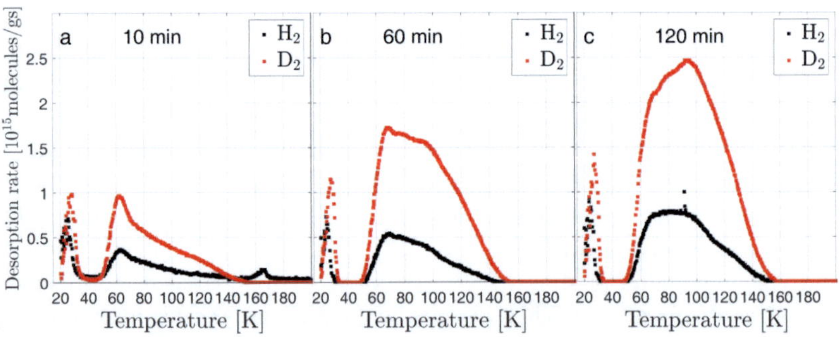

Figure 5.4: TDS spectra of IFP-4 exposed to 10 mbar of an equimolar isotope mixture at 50 K for different exposure times. (heating rate: $0.1\,K\,s^{-1}$)

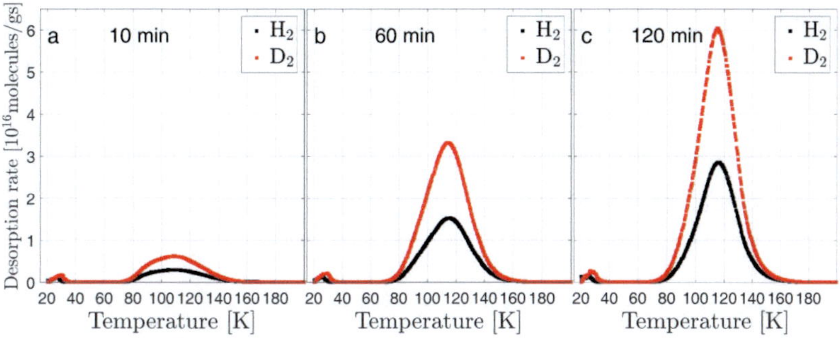

Figure 5.5: TDS spectra of IFP-4 exposed to 10 mbar of an equimolar isotope mixture at 77 K for different exposure times. (heating rate: $0.1\,\mathrm{K\,s^{-1}}$)

The maximum increases for longer exposure times for both materials, indicating that more isotopes are adsorbed during exposure. For IFP-4 the shape of the desorption peak differs between gas exposure at 50 K and 77 K. The amount of desorbed gas is over one order of magnitude higher for exposure at 77 K than at 50 K.

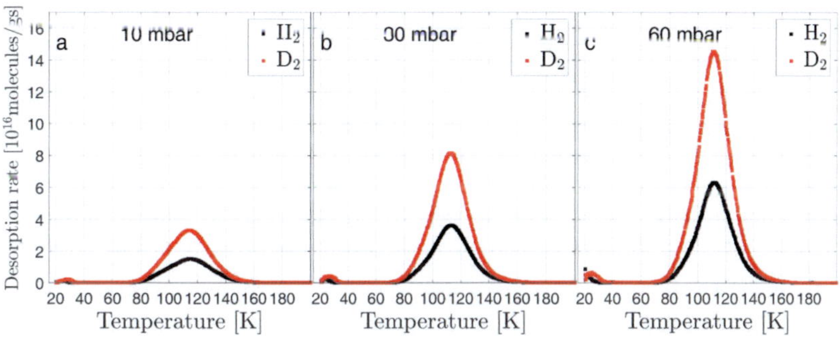

Figure 5.6: TDS spectra of IFP-4 exposed to different pressures of an equimolar isotope mixture at 77 K for an exposure time of 60 min. (heating rate: $0.1\,\mathrm{K\,s^{-1}}$)

Figure 5.6 depicts the behavior of the IFP-4 sample at 77 K exposure temperature by applying different loading pressures (10 mbar, 30 mbar, 60 mbar) for 60 min. Higher pressure leads for both isotopes to higher uptake. The maximum temperature does not change with loading pressure and also not with exposure time (figure 5.3, figure 5.5 and 5.6). Furthermore, after equimolar mixture exposure at 77 K all spectra show a nearly (within the experimental uncertainty) identical maximum temperature for H_2 and D_2.

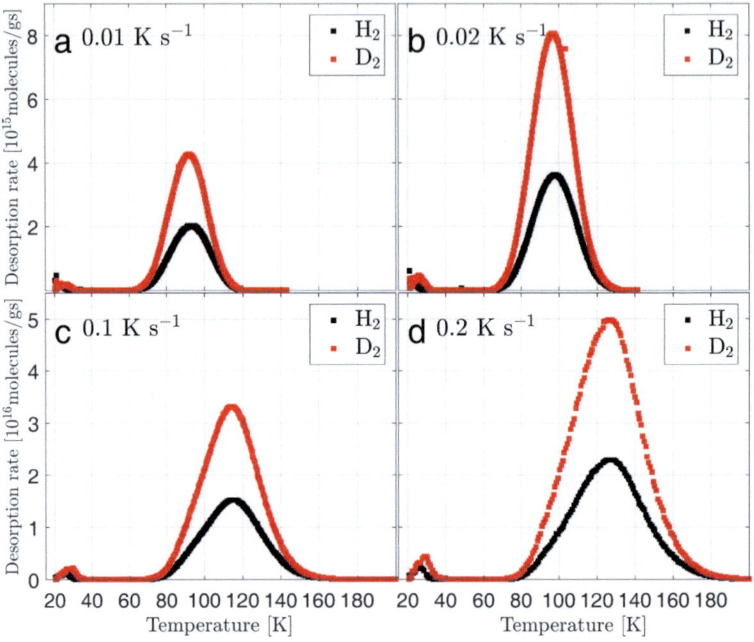

Figure 5.7: IFP-4 has been exposed to 10 mbar of an equimolar isotope mixture at 77 K for 60 min. For the desorption, different heating rates have been applied.

Different heating rates have been applied for IFP-4 exposed to an equimolar isotope mixture of 10 mbar at 77 K for 60 min. Figure 5.7 shows the thermal desorption spectra for different heating rates. Please note that the y-axes scales of the top and bottom row differ. For faster heating rates the maximum shifts to higher temperatures. Again, the maximum temperature for H_2 and D_2 are identical for all heating rates.

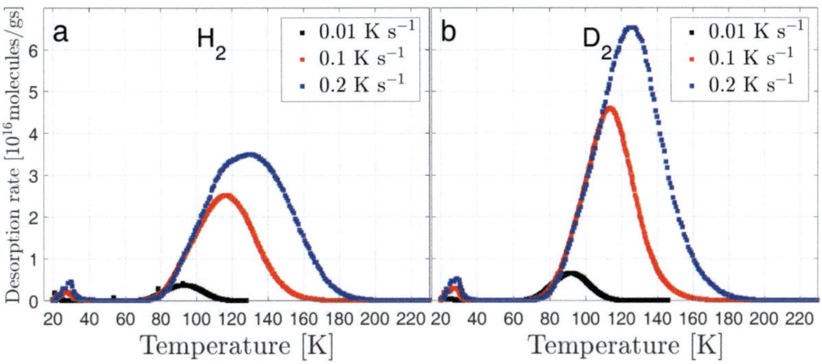

Figure 5.8: IFP-4 has been exposed to 10 mbar of H_2 (a) or D_2 (b) at 77 K for 60 min. For the desorption, different heating rates have been applied.

Different heating rates have been also applied after pure gas exposure and the spectra are presented in figure 5.8 versus the desorption temperature. This presentation of the desorption rate in molecules per gram and seconds on the y-axis yields smaller maxima for slower heating rates, however, if the TDS spectrum is plotted versus desorption time instead of temperature, the area under the curve (with unit molecules per gram) will be independent of heating rate. Again, the maxima shift to higher temperature for faster heating rates.

Measurements applying different heating rates allow the determination of the desorption energies for H_2 and D_2 (heating rate variation method chapter 2.3). The Kissinger plots presented in figure 5.9 have been generated using the temperature of the maximal desorption rate (T_{max}) in correlation with the corresponding heating rate (β) according to equation 2.20. The slope of the linear fit leads to the desorption energies, presented in table 5.1.

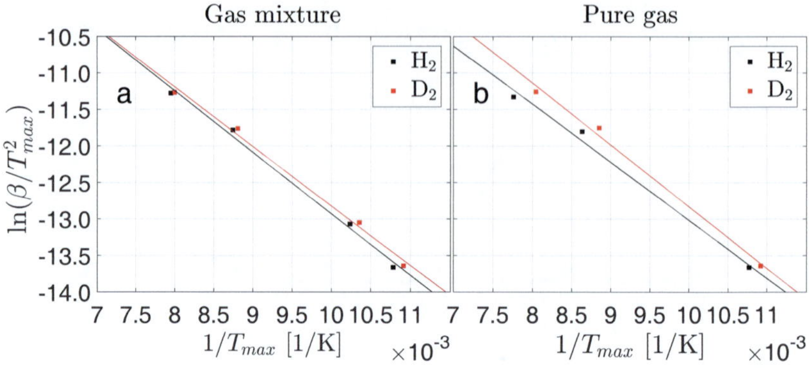

Figure 5.9: Kissinger plots for determining the desorption energy of hydrogen isotopes adsorbed to IFP-4. The slope of this linear fit contains the information of the desorption energy. The values have been taken from gas mixture measurements (a) and pure gas measurements (b).

	E_{des} in kJ mol^{-1}
$\mathbf{H_2}$ (mix)	7.0
$\mathbf{D_2}$ (mix)	6.8
$\mathbf{H_2}$ (pure)	6.6
$\mathbf{D_2}$ (pure)	7.0

Table 5.1: Desorption energy of H_2 and D_2 applied to IFP-4 as a mixture or as pure gas, determined with the Kissinger method.

Figure 5.10 shows the selectivity plotted against the exposure time for IFP-1 and IFP-3. The desorbed isotopes amount between 15 K and 300 K has been considered for the determination of the selectivity. IFP-1 exposed to 10 mbar (black) has no pronounced time dependence for the selectivity and is below $S = 2$, whereas for an exposure of 60 mbar (red), the selectivity increases from $S = 2.3$ (10 min) to $S = 3.7$ (90 min). For IFP-3 the selectivity first increases slightly from $S = 2.4$ (10 min) to $S = 2.8$ (30 min) and then decreases to $S = 2.3$ for 90 min exposure time. The corresponding TDS measurements are presented in figure 5.2.

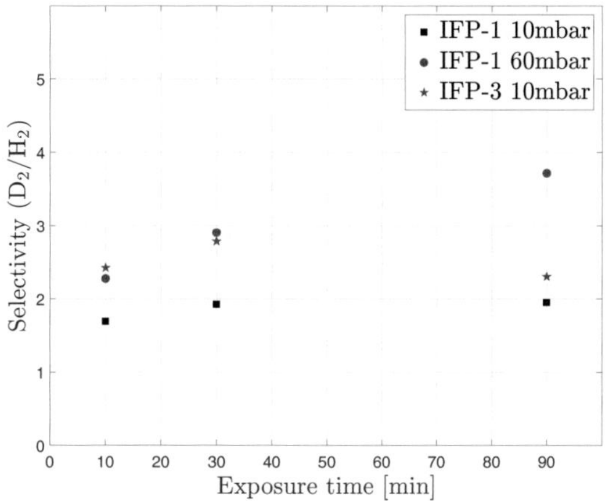

Figure 5.10: Selectivities of IFP-1 exposed to 10 mbar and 60 mbar at 30 K and of IFP-3 exposed to 10 mbar at 30 K for different exposure times.

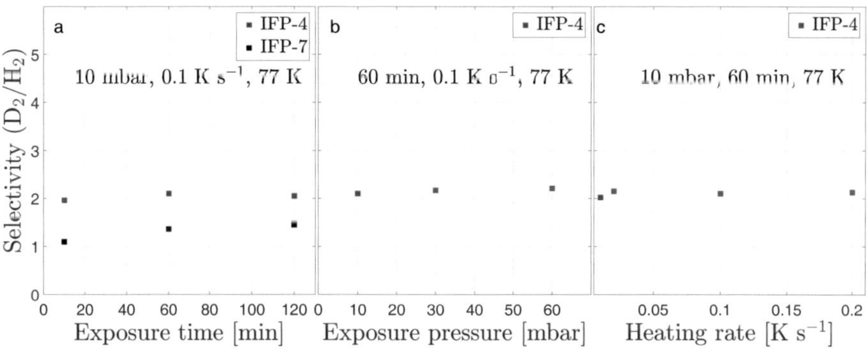

Figure 5.11: Selectivity of IFP-4 in dependence on exposure time, exposure pressure and heating rate.

Figure 5.11a presents the selectivity of IFP-7 and IFP-4 in dependence on exposure time. The desorbed isotopes amount between 15 K and 300 K has been considered for the determination of the selectivity. The selectivity of IFP-7 increases slightly for longer exposure times and reaches $S = 1.5$ at 120 min. IFP-4 shows a selectivity of $S \approx 2$ independent on exposure time, pressure and heating rate (figure 5.11).

5.1.2 Discussion

The IFP series is an isostructural MOF with different atoms or molecules at the aperture, which result in different channel/aperture sizes. Here, the influence of the aperture on Quantum Sieving has been studied on four materials IFP-1, -3, -7 and -4 with aperture sizes of 4.2 Å, 3.1 Å, 2.1 Å and 1.7 Å, respectively. These materials can be divided into materials with apertures larger and smaller than the kinetic diameter of hydrogen isotopes (2.89 Å). In this section first the adsorption of hydrogen isotopes and then their separation is discussed.

Hydrogen isotope adsorption

In table 5.2, the adsorbed amounts of the IFP samples loaded with either pure H_2 or pure D_2 are given, together with the respective temperature of the maximum. The corresponding TDS spectra are presented in figure 5.1.

(pore aperture)	IFP-1 (4.2 Å)	IFP-3 (3.1 Å)	IFP-7 (2.1 Å)	IFP-4 (1.7 Å)
H_2				
T_{max} [K]	51	57	116	124
uptake [molecules/g]	4.0×10^{21}	1.2×10^{21}	3.2×10^{19}	6.1×10^{18}
D_2				
T_{max} [K]	53	54	107	118
uptake [molecules/g]	5.6×10^{21}	1.5×10^{21}	3.0×10^{19}	7.9×10^{18}

Table 5.2: For the samples IPF-1, -3, -7 and -4, the temperature of maximum desorption and the total gas amount desorbed in this maximum (uptake) is shown. For corresponding TDS spectra see figure 5.1.

<u>IFP-1 and IFP-3</u>

IFP-1, the sample with the largest aperture size adsorbs the highest amount of isotopes, but its desorption maximum is also at the lowest desorption temperature (see table 5.2). The desorption starts at 40 K with a steep increase in desorption rate and a high amount of gas is released immediately.

The pure gas measurement of IFP-1 presented in figure 5.1 reveals an H_2 uptake of 4.0×10^{21} molecules g^{-1} which is equivalent to 1.35 wt%. This is in accordance with Debatin *et al.*, which reported an H_2 uptake of 1.5 wt% from isotherm measurements of IFP-1. [16].
The adsorbed amount is almost identical for loading times in the range from 10 min to 90 min (figure 5.2), which indicates that already after 10 min all adsorption sites can be reached by the hydrogen isotopes. For IFP-3 with an aperture size of 3.1 Å, the total adsorbed amount increases slightly for longer exposure times, which implies a small diffusion limitation.

IFP-7 and IFP-4

IFP-7 and IFP-4 adsorb less gas compared to the other two materials and their desorption maximum rises at much higher temperatures, namely above 100 K (see table 5.2). These materials show a remarkable increase in uptake for longer exposure times and higher exposure pressures (see figures 5.3, 5.4, 5.5, 5.6). The observed dependence on exposure time and pressure reveals that the diffusion is hindered and slower, therefore, higher exposure times are chosen.
This directly reflects the small aperture size of IPF-7, which is restricted by methoxy groups to a diameter of 2.1 Å and the small aperture of IFP-4, which is formed by ethyl groups resulting in a diameter of 1.7 Å. Thus, assuming a rigid structure, hydrogen isotopes with a kinetic diameter of 2.89 Å, can not penetrate into the materials. [100]
Adsorption measurements presented in literature have shown that at 77 K no penetration of N_2 (kinetic diameter: 3.6 Å) into IFP-4 is possible. The low value of the BET-surface area of 5 m^2g^{-1} measured at 77 K leads to the conclusion that the molecules populated only the outer surface of the material, but at 273 K N_2 can penetrate the structure. This points towards a thermally activated flexibility of the ethyl groups. [15]
Hydrogen isotopes possess a smaller kinetic diameter than N_2 and can therefore enter the structure at lower temperatures. In this work, the temperature dependent opening of the apertures of IFP-7 has been investigated with different temperatures. While at 30 K exposure temperature, no penetration of hydrogen isotopes into the pore channels occurs, at 40 K a maximum appears in the TDS spectrum, which reveals that penetration into the structure becomes possible. For IFP-4, at 50 K exposure, a very low adsorption has been observed, which shows that at this temperature, the aperture starts to open. Increasing the exposure temperature to 77 K leads to a tenfold higher hydrogen isotope uptake (figure 5.4 and figure 5.5).

Desorption energy of hydrogen isotopes adsorbed in IFP-4

For IFP-4 the desorption energy of H_2 has been determined in this work with TDS to a value of 7.0 kJ mol^{-1} (from mixture measurements) and 6.6 kJ mol^{-1} from pure H_2 measurements. This matches very well the value from Denysenko of 6.9 kJ mol^{-1} [93], which has been obtained with isotherm measurements. The desorption energies of D_2 have also been determined with TDS and are similar to H_2 (see table 5.1).

Separation of hydrogen isotopes

IFP-1 and IFP-3

For the separation of isotopes, Quantum Sieving is suitable only when the pore aperture is slightly bigger than the kinetic diameter of the isotopes to be separated (see chapter 2.2.2). The pore apertures of IFP-1 (4.2 Å) and IFP-3 (3.1 Å) fulfill this requirement very well for hydrogen isotopes, which have a kinetic diameter of 2.89 Å.

(pore aperture)	IFP-1 (4.2 Å)	IFP-3 (3.1 Å)	IFP-7 (2.1 Å)	IFP-4 (1.7 Å)
$\mathbf{T}_{exp}$ **[K]**	30	30	77	77
$\mathbf{S}_{max}$ (10 mbar)	2.0 (30 min)	2.8 (30 min)	1.5 (120 min)	2.1 (60 min)
$\mathbf{S}_{max}$ (60 mbar)	3.7 (90 min)	–	–	2.2 (60 min)

Table 5.3: Highest obtained selectivity after an equimolar mixture has been loaded to IFP-1, IFP-3, IFP-7 and IFP-4.

Comparison of IFP-1 and IFP-3 measured under identical experimental conditions (10 mbar, 30 K) shows that IFP-3 with the smaller pore aperture has the higher selectivity (see figure 5.10 and table 5.3). For the optimal separation of isotopes, the parameters used for the exposure (e.g. pressure, time and temperature) are crucial. The highest selectivity of $S = 3.7$ has been obtained by exposing IFP-1 for 90 min to an equimolar mixture of 60 mbar at 30 K. Quantum Sieving is based on the diffusion limitation of the lighter isotope. Owing to the small exposure time dependence of the total adsorbed amount (see figure

5.2), basically no diffusion limitation occurs for IFP-1 and IFP-3 and, therefore, the observed sieving of the isotopes is mainly caused by Chemical Affinity Sieving occurring from the adsorption on the pore walls (see chapter 2.2.3).
The optimum pore diameter for Quantum Sieving has been determined to be between 3.0 Å and 3.4 Å by Oh *et al.* [101]. From the material series investigated in this work, the aperture diameter of IFP-3 lies in this range, however, as it can be seen from table 5.3 the maximum selectivity is rather low. The flexibility of the aperture observed for IFP-4 and IFP-7 occurs as well for IFP-3, which, even at 30 K, may lead to an aperture which exceeds the range of the optimum pore diameter.

IFP-7 and IFP-4

The small pore apertures of IFP-7 and IFP-4 have a strong influence on diffusion of gas into the framework. IFP-7 shows a very low selectivity, which is below $S = 2$ (see figure 5.11 and table 5.3). IFP-4 exposed at 77 K has a constant value of $S \approx 2$ and no exposure time, pressure or heating rate dependence. The heating rate has no impact on the selectivity, these values are shown to further substantiate the constancy of the value.
Quantum Sieving has been first investigated experimentally by Teufel *et al.* with equimolar hydrogen isotope mixtures in Cl-MFU-4. [102] As IFP-4, this material has flexible apertures. The reported selectivity of 7.5 for Cl-MFU-4 (at 60 K, 10 mbar, 15 min) is much higher than the selectivity of IFP-4. This is ascribed to significant differences in the structure of the two materials. In contrast to Cl-MFU-4, for IFP-4 at 77 K the selectivity remains at an approximately constant value of $S = 2$ for different exposure times or pressures. The small channels lead to a chain-like filling and no passing between the isotopes in the channels is possible. Therefore, the desorption after exposure to an equimolar mixture shows identical maximum temperatures for both isotopes, whereas the maximum of pure gas desorption appears for D_2 at lower temperatures than for pure H_2. This means that Quantum Sieving occurs only at the outermost pore aperture and no further exchange is possible inside the channels. In contrast, the Cl-MFU-4 structure consists of a combination of large and small pores, thereby penetrated molecules can be stored in the large pore. Each further penetration through the small apertures into the next larger pore leads to an additional sieving and, therefore, an enhancement of the selectivity.

Summary
The separation of hydrogen isotopes has been investigated with selected MOFs from the IFP series, whose different aperture diameters (1.7 Å-4.2 Å) meet the size requirement for Quantum Sieving. The aperture diameters directly influence the desorption behavior, namely, for smaller diameters the diffusion of the isotopes out of the material is hindered and thus slower. Therefore, the maxima appear at higher temperatures in the spectrum (see table 5.2).
IFP-7 (2.1 Å) and IFP-4 (1.7 Å) have the smallest aperture diameter of the chosen IFP samples, which would not allow adsorption of hydrogen molecules (2.89 Å) from a geometrical point of view. As expected from N_2 adsorption measurements, the apertures show a temperature dependent dynamic opening and, therefore, at a certain material specific temperature, the molecules can enter the pores.
For IFP-4, the material with the smallest pore aperture, the selectivity is approximately $S \approx 2$ and shows no dependence on exposure time or pressure. In the narrow pore channels of the IFP series hydrogen isotopes cannot pass each other once they are adsorbed. Therefore, the selective adsorption occurs only at the outermost pore aperture and no further enhancement of the selectivity is possible in the material by varying the exposure parameters. The highest selectivity reached with the IFP samples is $S = 3.7$ for IFP-1 at 30 K, which is only half of the up to now highest known value of $S = 7.5$ for Quantum Sieving in Cl-MFU-4 at 60 K [102].

5.2 Helium adsorption in IFP-1, IFP-3 and IFP-4

5.2.1 Results

After activation, the IFP-samples were exposed to a helium atmosphere and the non-adsorbed gas has been evacuated at base temperature for IFP-1 and IFP-3 and at exposure temperature for IFP-4. The measured desorption rate is divided by the sample mass to allow a quantitative comparison between the different samples.
Figure 5.12a shows the TDS spectra of IFP-1 exposed to 10 mbar ^{4}He at 20 K for different exposure times. A sharp peak occurs at 22 K, which increases for longer exposure times. The TDS spectra of IFP-3 exposed to 10 mbar of ^{4}He at 30 K exposure temperature for different exposure times is depicted in figure 5.12b. A maximum is observed at 28 K. In contrast to IFP-1, the maximum of IFP-3 becomes smaller for longer exposure times.

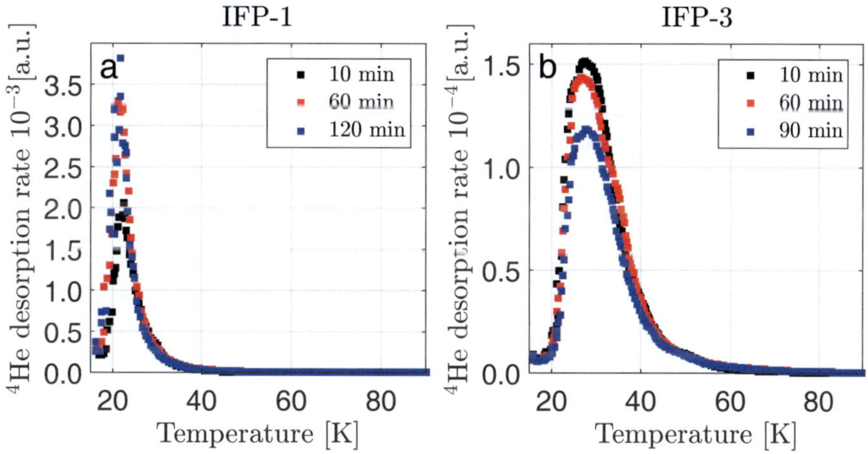

Figure 5.12: a) TDS spectra of IFP-1 exposed at 20 K to 10 mbar ^{4}He for different exposure times. b) TDS spectra of IFP-3 exposed at 30 K to 10 mbar ^{4}He for different exposure times. (heating rate: $0.1\,\mathrm{K\,s^{-1}}$)

TDS spectra of IFP-1 and IFP-3 have been recorded after exposure at 20 K to ^{4}He pressures from 1 mbar to 900 mbar for 10 min. Figure 5.13 shows as an example the spectra obtained by exposing each material to 10 mbar. The colored area depicts the amount of desorbed ^{4}He after exposure at 20 K for 10 min and evacuation of the non-adsorbed gas at base temperature. Since the desorption rate at base temperature is low, the amount of desorbed atoms (blue area) can be set equal to the amount of adsorbed atoms at exposure temperature. Therefore, plotting the adsorbed amount versus the exposure pressure leads to an adsorption isotherm at exposure temperature (figure 5.14). The isotherms have a steep increase at low pressures and start to saturate for higher pressures, which is similar to a type I isotherm [103].

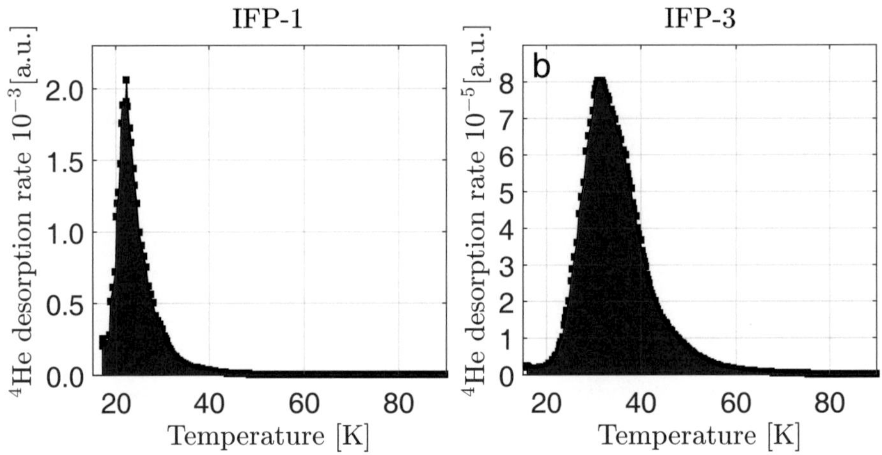

Figure 5.13: TDS spectra after exposure of a) IFP-1 b) IFP-3 at 20 K to 10 mbar of ^{4}He for 10 min. The colored area represents the desorbed gas amount. (heating rate: 0.1 K s^{-1})

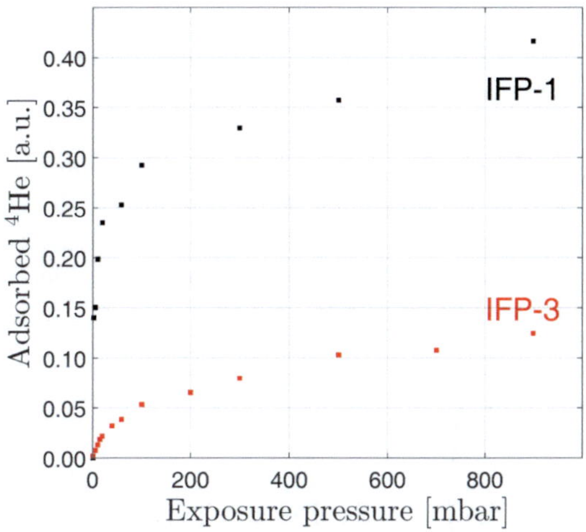

Figure 5.14: ^{4}He adsorption for IFP-1 and IFP-3 exposed at 20 K for 10 min. The exposure pressure has been varied from 1 mbar to 900 mbar.

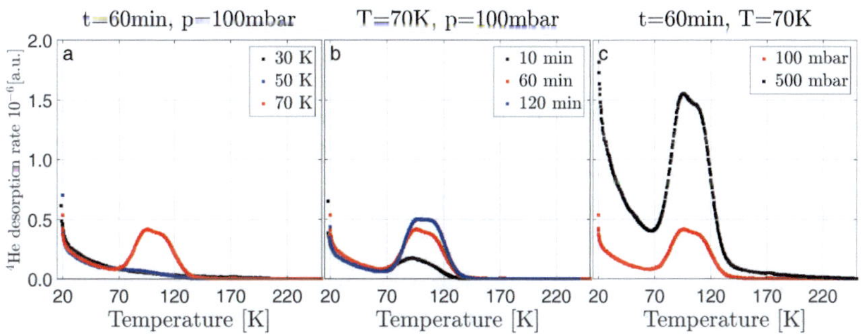

Figure 5.15: TDS spectra of IFP-4 a) for different exposure temperatures exposed to 100 mbar ^{4}He for 60 min b) for different exposure times exposed to 100 mbar ^{4}He at 70 K c) for different ^{4}He exposure pressures exposed at 70 K for 60 min. (heating rate: $0.1\,\mathrm{K\,s^{-1}}$)

Figure 5.15a shows the desorption spectra of IFP-4 exposed to 100 mbar ^{4}He for 60 min at different exposure temperatures. After exposure at 30 K no desorption maximum occurs, the desorption rate decreases monotonically with temperature. A very small maximum rises for exposure at 50 K and for exposure at 70 K a clearly defined desorption maximum appears at 100 K desorption temperature. For longer exposure times at 70 K exposure temperature this maximum increases, which shows that more atoms penetrate into the material and get adsorbed (figure 5.15b). The desorption signal up to 75 K desorption temperature, i.e. below the maximum, remains constant even for longer exposure times. Increasing the exposure pressure, leads to an overall increase of the desorption signal after exposure at 70 K (figure 5.15c).

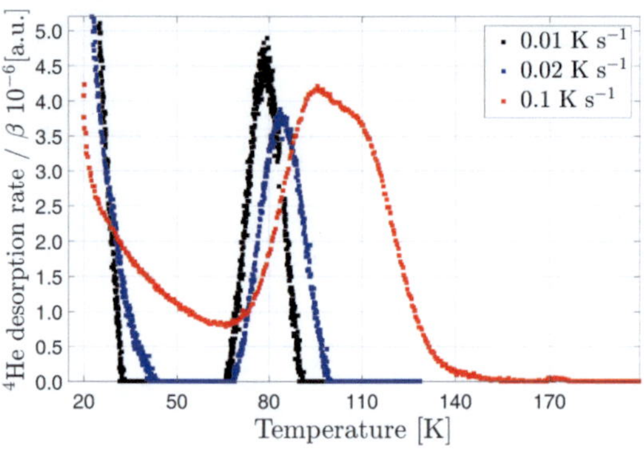

Figure 5.16: 100 mbar ^{4}He have been exposed to IFP-4 at 70 K for 60 min. The desorption process has been performed with different heating rates. The measured desorption rate has been divided by the heating rate β in order to remove the different durations of the measurements for different heating rates.

TDS spectra of IFP-4 exposed to 100 mbar ^{4}He at 70 K exposure temperature for 60 min have been measured with various heating rates (see figure 5.16). For higher heating rates, the desorption maximum shifts to higher desorption temperatures.

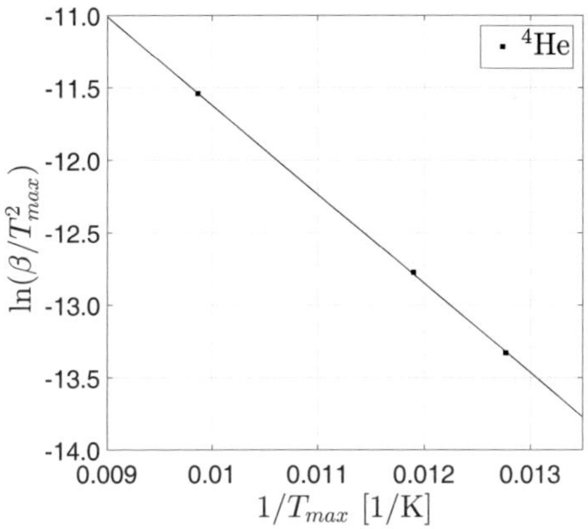

Figure 5.17: Kissinger plot for the determination of the desorption energy of ^{4}He adsorbed in IFP-4. β is the heating rate and T_{max} is the temperature of the maximum desorption rate.

The desorption energy of ^{4}He has been determined with the Kissinger method (see chapter 2.3). This method has been already applied for IFP-4 exposed to hydrogen isotopes (chapter 5.1.2). For ^{4}He a desorption energy of 5.1 kJ mol^{-1} has been determined from the slope of the linear fit presented in figure 5.17.

5.2.2 Discussion

The comparable kinetic diameters of hydrogen (2.89 Å [100]) and helium (2.60 Å [104]), gave the initiative to investigate the adsorption of helium in the small pore channels of the IFP samples. Since the mass spectrometer signal cannot be calibrated as it is possible for hydrogen isotopes (see chapter 3.1.2), for a quantitative comparison between the samples, the desorption rate is divided by the sample mass.

Helium-4 adsorption

IFP-1 and IFP-3

As already observed for hydrogen isotopes, the TDS spectra of IFP-1 and -3 show maxima at relatively low temperatures, whereas the desorption maximum for IFP-4 occurs at higher temperatures. The TDS maxima of IFP-1 and -3 after exposure to hydrogen isotopes are at higher desorption temperatures, compared to the TDS maxima after exposure to helium. The desorption rate of hydrogen isotopes adsorbed in IFP-1 starts to increase around 40 K and a broad peak develops (see figure 5.1a). However, for IFP-1 exposed to helium, a very sharp desorption peak rises already around 15 K. Thus, compared to hydrogen isotopes, the desorption energy of helium is lower and no diffusion limitation occurs for the desorption process.
Due to the smaller pore aperture of IFP-3 (3.1 Å), the pore walls have a higher curvature and, therefore, a stronger adsorption potential (see chapter 2.1) compared to IFP-1. The IFP-3 maximum is broader (between 20 K and 60 K desorption temperature), whereas the maximum for IFP-1 ends around 40 K indicating that the helium atoms are weaker bound than in IFP-3.
The reduced helium adsorption for longer exposure times has only been observed for IFP-3, which has a smaller pore aperture than IFP-1 but a larger one than IFP-4. Therefore, the smaller pore aperture cannot be the reason for lower helium adsorption for longer exposure times. An approach is to attribute this observation to the interaction between the ^{4}He atoms and the bromine substituents of IFP-3, which restrict the pore aperture. A decreasing helium adsorption for longer exposure times has also been observed in a bromine containing material, which will be discussed in the next chapter.

IFP-4

The pore aperture of IFP-4 is 1.7 Å for a stiff structure model. Therefore, penetration of ^{4}He can only occur due to an opening of the apertures for higher temperatures. After exposing IFP-4 at 30 K to ^{4}He, no maximum occurs in the TDS spectrum (see figure 5.15a). At 50 K exposure temperature a maximum appears, which is especially visible if the desorption rate is plotted logarithmically (see figure 5.18).

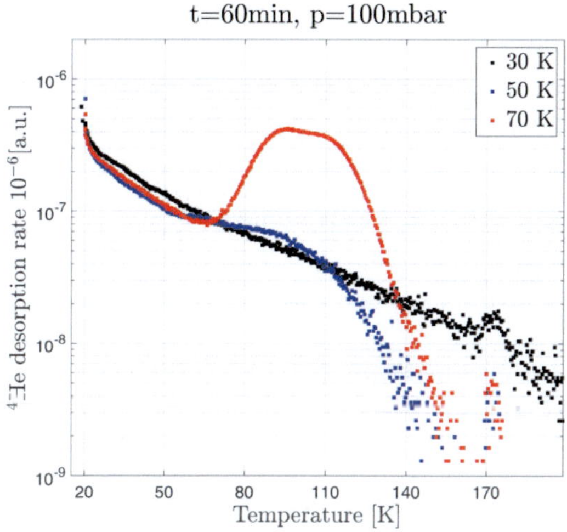

Figure 5.18: TDS spectra of IFP-4 recorded after exposing to 100 mbar ^{4}He for 60 min at different exposure temperatures, plotted logarithmically. (heating rate: $0.1\,\mathrm{K\,s^{-1}}$)

A higher desorption maximum develops if the material is exposed at 70 K. Longer exposure times at 70 K lead to a higher maximum, but not to a higher signal below 75 K desorption temperature, whereas for higher exposure pressures both signals increase (see figure 5.15b, c). Hence, the increasing maximum for higher exposure time shows that the diffusion into the material depends on time, like it has been observed for hydrogen isotopes. Due to the time independence of the signal at low temperatures (below 75 K) it is most probably, that this signal occurs from helium atoms adsorbed in the inter-particle voids of the powder and not from penetration into the material. The increased signal for higher pressures shows that more helium atoms are allocated in these voids (see figure 5.15c).

Desorption energy of IFP-4

With the Kissinger method (see chapter 2.3 and figure 5.17) a desorption energy of 5.1 kJ mol^{-1} has been determined for ^{4}He in IFP-4. For hydrogen isotopes adsorbed in IFP-4 a higher desorption energy has been obtained (between 6.6 kJ mol^{-1} and 7.0 kJ mol^{-1} see table 5.1). Therefore, helium with a smaller kinetic diameter releases easier from the pore structure.

Summary

The adsorption of ^{4}He has been investigated in three MOFs of the IFP series, which have different aperture sizes in the range of 1.7 Å to 4.2 Å. These are the first measurements of ^{4}He adsorption in MOFs with TDS. For the materials with larger pores (IFP-1 and IFP-3) TDS spectra have been recorded at different applied pressures. The obtained data could be plotted as type I isotherms, which confirms that low temperature TDS can be used to study helium adsorption.

Compared to hydrogen isotopes (2.89 Å), ^{4}He has a smaller kinetic diameter of 2.60 Å and can diffuse faster out of the material. Furthermore, the desorption of ^{4}He appears at lower temperatures. For the material with the highest desorption temperature, IFP-4, the desorption energy has been determined for ^{4}He (5.1 kJ mol^{-1}) and hydrogen isotopes (between 6.6 kJ mol^{-1} and 7.0 kJ mol^{-1}), which shows quantitatively the weaker adsorption of ^{4}He due to its weaker van der Waals interaction with the pores. From a geometrical point of view ^{4}He should not be able to penetrate into IFP-4 at all, but due to the flexibility of the aperture, the gate opening effect occurs and adsorption is possible for exposure at 50 K and higher temperatures.

The low selectivities obtained for hydrogen isotopes set a substantially lower expectation level on the selectivities obtainable with helium isotopes. Thus, no selectivity measurements with the precious and expensive ^{3}He were carried out with the IFP samples.

5.3 Helium isotope adsorption and separation in Cl-MFU-4 and Br-MFU-4

The high selectivity of $S = 7.5$ [102] obtained for hydrogen isotopes in Cl-MFU-4 motivated to investigate the separation of helium isotopes in this material. Due to the smaller kinetic diameter of helium isotopes of 2.60 Å compared to the one for hydrogen isotopes 2.89 Å also Br-MFU-4 with smaller aperture sizes has been considered for these investigations.

5.3.1 Results

Cl-MFU-4 has been activated at 423 K for 60 min in the TDS chamber under high vacuum before the measurements series started. Compared to Cl-MFU-4 a higher temperature of 500 K and a longer activation time of 150 min has been chosen for Br-MFU-4. After exposure to helium atmosphere, the evacuation of the non-adsorbed gas has been performed at base temperature (see chapter 3.1.3) if it is not mentioned explicitly otherwise. In order to compare the spectra of the different investigated samples, the diagrams show the measured desorption rate divided by the sample mass.

Cl-MFU-4

Cl-MFU-4 has been exposed at 30 K and 77 K to 10 mbar ^{4}He for different exposure times. The TDS spectra show one maximum at around 40 K and 50 K after exposure at 30 K and 77 K, respectively. The desorption maximum increases for longer exposure time at 30 K (figure 5.19a), whereas at 77 K (figure 5.19b), the desorption maximum is constant for different exposure times (table 5.4). For exposure at 30 K the desorption maximum is much higher, showing that more ^{4}He is adsorbed than for exposure at 77 K.

$\mathbf{t}_{ex}$ **[min]**	$\mathbf{T}_{ex} = 30\,\mathrm{K}$	$\mathbf{T}_{ex} = 77\,\mathrm{K}$
10	7.0×10^{-06}	2.6×10^{-06}
60	16.9×10^{-06}	2.7×10^{-06}
120	19.7×10^{-06}	2.7×10^{-06}

Table 5.4: Adsorbed amount of ^{4}He at 30 K and 77 K exposure temperature for different exposure times. The corresponding TDS spectra are shown in figure 5.19.

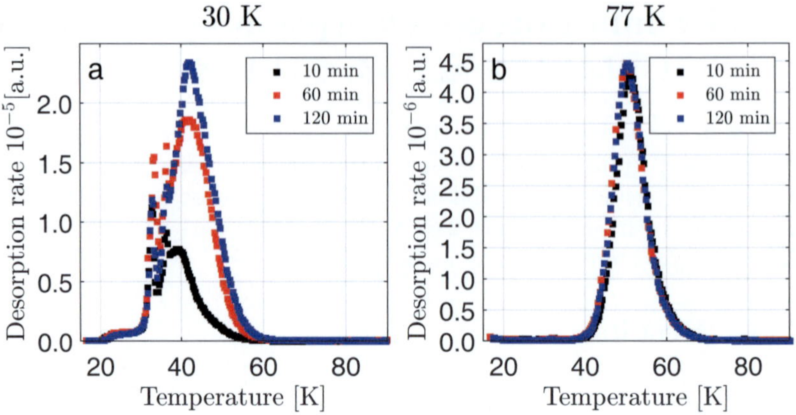

Figure 5.19: TDS spectra of Cl-MFU-4 exposed to 10 mbar of ^{4}He at 30 K (a) and 77 K (b) for different exposure times. Please note the different scaling on the y-axes. (heating rate: $0.1\,\mathrm{K\,s^{-1}}$)

Figure 5.20 shows the TDS spectra after exposure to 10 mbar ^{4}He for 60 min at different exposure temperatures. For exposure temperatures between 30 K and 60 K the adsorbed amount decreases with increasing exposure temperature (figure 5.20a), whereas it stays constant for higher exposure temperatures (figure 5.20b). The maximum temperature shifts to higher values between 30 K and 60 K exposure temperature (figure 5.20a). For exposure temperatures above 60 K the maximum is at slightly lower temperatures and remains constant (figure 5.20b). Please take note that after exposing the sample to the gas at a certain temperature, the sample has been cooled under gas atmosphere to base temperature and then the non-adsorbed gas has been evacuated.

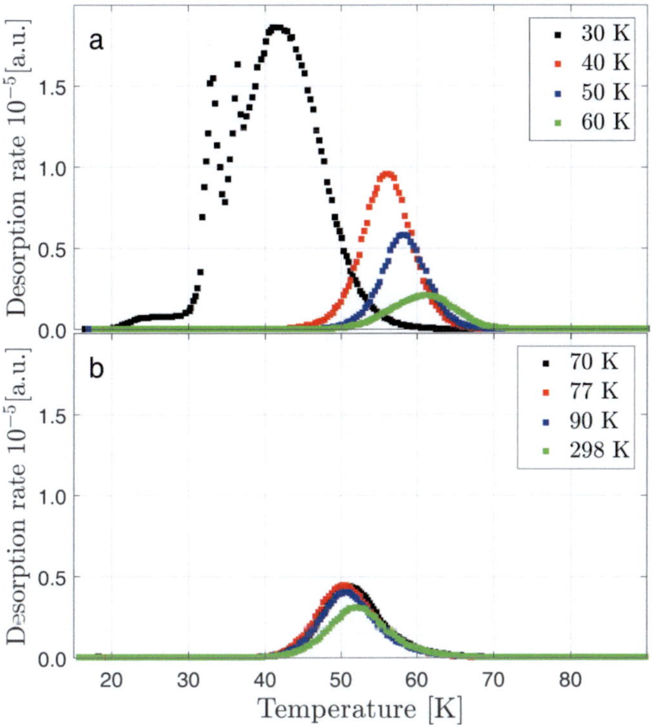

Figure 5.20: TDS spectra of Cl-MFU-4 exposed to 10 mbar ^{4}He for 60 min at different exposure temperatures. (heating rate: $0.1\,\mathrm{K\,s^{-1}}$)

Figure 5.21 shows the TDS spectra after exposure to 10 mbar ^{4}He for 60 min at 40 K and 77 K . The evacuation of the non-adsorbed gas has been performed at exposure temperature and at base temperature (here, 15 K). The maxima at 40 K exposure temperature are similar, whereas no maximum occurrs in the spectrum if evacuation has been done at 77 K exposure temperature.

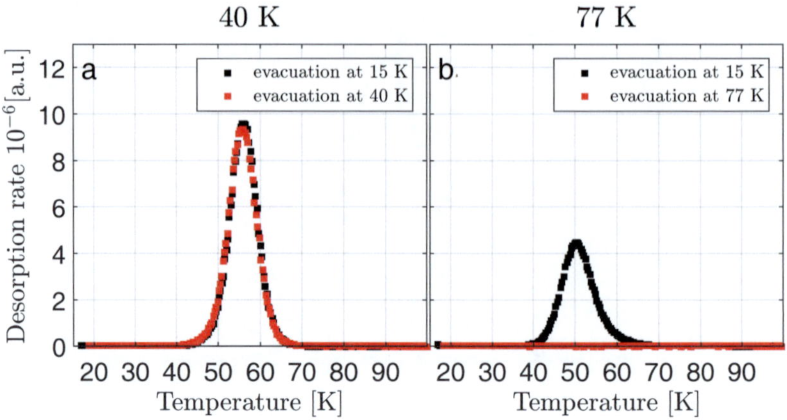

Figure 5.21: 10 mbar of ^{4}He have been exposed for 60 min at 40 K (a) and 77 K (b). The evacuation of the non-adsorbed gas occurred at exposure temperature and at base temperature (here 15 K). (heating rate: $0.1\,\mathrm{K\,s^{-1}}$)

An equimolar mixture of ^{3}He and ^{4}He was prepared in the prechamber for further experiments. Owing to the low pressure of the ^{3}He reservoir it was difficult to adjust the pressure and, therefore, the final pressure of the mixtures varies by a few millibar. The desorption spectra of Cl-MFU-4 exposed to equimolar mixtures of helium isotopes at different exposure temperatures for 10 min is presented in figure 5.22. The amount of adsorbed ^{4}He is always higher than the amount of adsorbed ^{3}He. With increasing exposure temperature, the adsorbed amount decreases for both isotopes. The maximum temperature of both isotopes shifts to higher values for higher exposure temperatures.

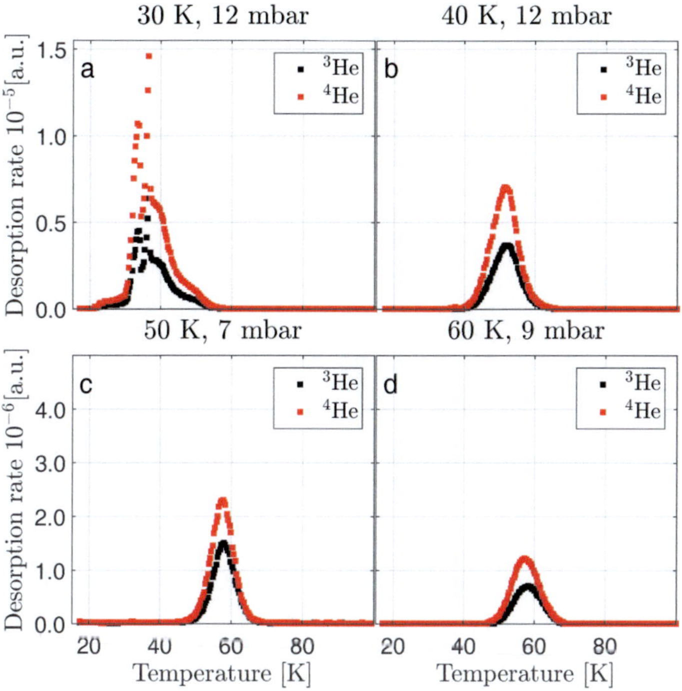

Figure 5.22: TDS spectra of Cl-MFU-4 exposed to an equimolar mixture of ^{3}He and ^{4}He for 10 min at different exposure temperatures. Please note the different scalings of the y-axes. (heating rate: $0.1\,\mathrm{K\,s^{-1}}$)

The TDS spectra after exposure to an equimolar helium isotope mixture of 9 mbar and 55 mbar at 60 K exposure temperature for 10 min exposure time are shown in figure 5.23. The amount of adsorbed helium isotopes increases with pressure. Both isotope maxima shift to higher temperatures for higher pressure (around 58 K for 9 mbar and around 61 K for 55 mbar).

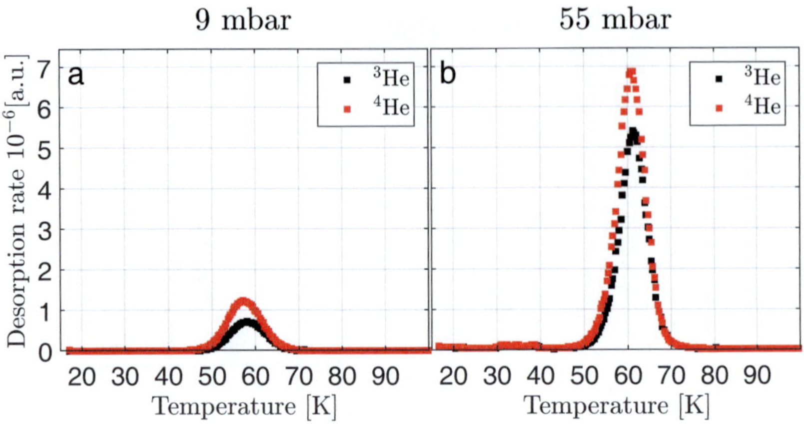

Figure 5.23: 9 mbar (a), 55 mbar (b) of an equimolar mixture of ^{3}He and ^{4}He has been exposed to Cl-MFU-4 for 10 min at 60 K. (heating rate: $0.1\,\mathrm{K\,s^{-1}}$)

exposure temperature	**30 K**	**40 K**	**50 K**	**60 K**	**77 K**
time [min]	10	10	10	10	
pressure [mbar]	12	12	7	9	
selectivity	2.19	1.84	1.45	1.80	
time [min]	120	5			120
pressure [mbar]	17	6			20
selectivity	1.94	1.72			1.41
time [min]	10	5			
pressure [mbar]	17	18			
selectivity	2.25	1.76			
time [min]	10			10	
pressure [mbar]	45			55	
selectivity	2.11			1.3	

Table 5.5: Selectivity of Cl-MFU-4 after applying an equimolar ^{3}He and ^{4}He mixture for varying parameters.

Table 5.5 gives an overview of the applied parameters and the resulting selectivities of Cl-MFU-4 after an equimolar mixture of ^{3}He and ^{4}He has been applied. The amount of desorbed isotopes between 15 K and 300 K has been considered for the determination of the selectivity. These values have been calculated under the assumption that the mass spectrometer has the same sensitivity for mass three and mass four.

Br-MFU-4

Exposure of Br-MFU-4 to 15 mbar of pure ^{3}He or ^{4}He at 298 K for 60 min leads to the desorption spectra presented in figure 5.24. For both isotopes two maxima are observed one at lower and one at higher temperature (see table 5.6). For ^{3}He, the adsorbed amount at lower temperatures exceeds the adsorbed amount at higher temperatures and for ^{4}He the adsorbed amount at higher temperatures exceeds the amount at lower temperatures. The areas of the maxima are given in table 5.6 in parentheses.

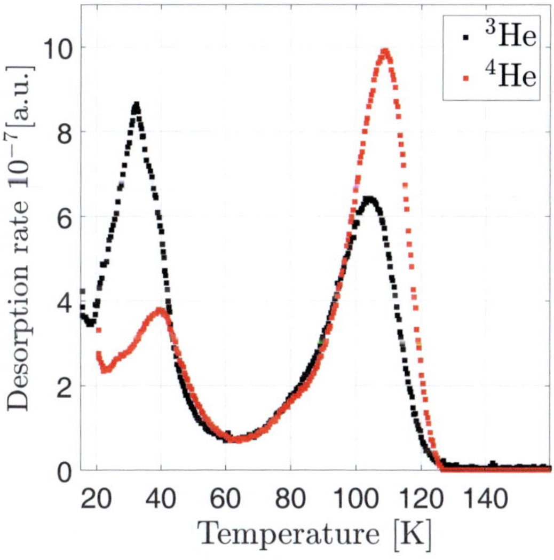

Figure 5.24: 15 mbar of pure ^{3}He or ^{4}He have been exposed at room temperature to Br-MFU-4 for 60 min. (heating rate: 0.1 K s^{-1})

	$\mathbf{T}_{max}$ **[K]** (area 10^{-06})	$\mathbf{T}_{max}$ **[K]** (area 10^{-06})
^{3}**He**	32.5 (1.14)	104.4 (1.01)
^{4}**He**	40.0 (0.63)	108.6 (1.37)

Table 5.6: Temperature of the desorption maxima of Br-MFU-4 for ^{3}He and ^{4}He. The corresponding TDS spectra are shown in figure 5.24.

Varying the exposure time leads to the TDS spectra presented in figure 5.25 (exposure pressure 10 mbar, exposure temperature 77 K). The low temperature maximum decreases for both isotopes with longer exposure time. The high temperature desorption maximum, first increases from 10 min to 60 min exposure time and remains constant for longer times.

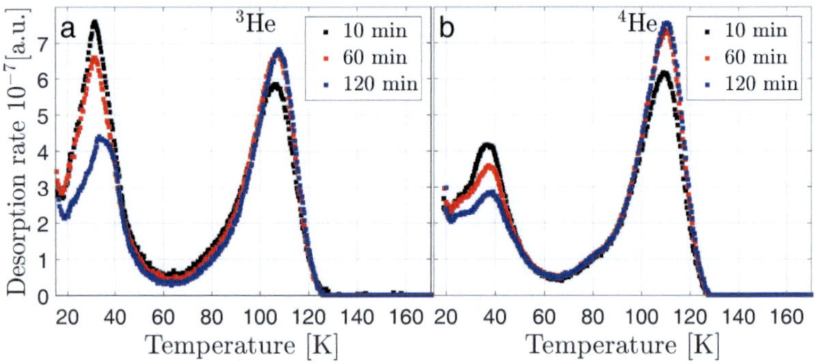

Figure 5.25: 10 mbar of ^{3}He (a) or ^{4}He (b) have been exposed to Br-MFU-4 at 77 K for different exposure times. (heating rate: 0.1 K s^{-1})

Figure 5.26 shows TDS spectra of Br-MFU-4 exposed to 10 mbar of ^{3}He or ^{4}He for 60 min at different exposure temperatures. The low temperature maximum does not change up to 50 K exposure temperature. The high temperature maximum first increases continuously up to 50 K and then decreases for higher exposure temperatures.

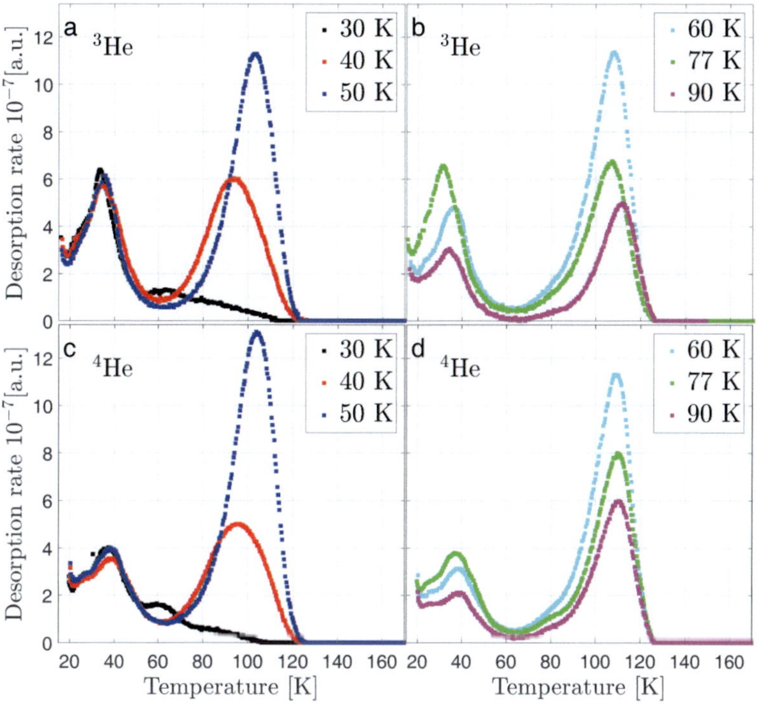

Figure 5.26: TDS spectra of Br-MFU-4 exposed to 10 mbar ^{3}He or ^{4}He for 60 min at different exposure temperatures. (heating rate: $0.1\,\mathrm{K\,s^{-1}}$)

Different heating rates have been applied to Br-MFU-4, which was exposed to 10 mbar of ^{3}He or ^{4}He for 60 min at 77 K. The TDS spectra are presented in figure 5.27, indicating a clear shift of the desorption maximum to lower temperatures for slower heating rates.
The Kissinger plots presented in figure 5.28 have been generated using the temperature of the maximal desorption rate (T_{max}) together with the corresponding heating rate (β). According to equation 2.20 the slope of the linear fit leads to the desorption energies of $1.6\,\mathrm{kJ\,mol^{-1}}$ for ^{3}He and $1.5\,\mathrm{kJ\,mol^{-1}}$ for ^{4}He. The almost identical desorption energies are as expected owing to the very similar TDS spectra of these isotopes for different heating rates (figure 5.27).

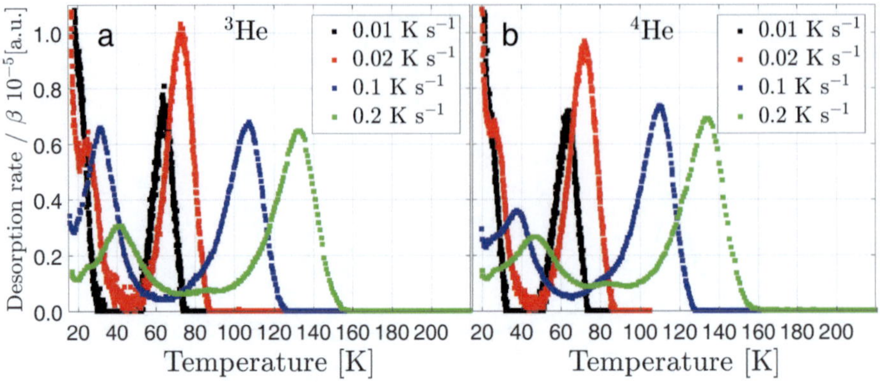

Figure 5.27: 10 mbar of ^{3}He (a) or ^{4}He (b) have been exposed to Br-MFU-4 at 77 K for 60 min. For the desorption process various heating rates have been applied. The measured desorption rate has been divided by the heating rate β in order to remove the different durations of the measurements for different heating rates.

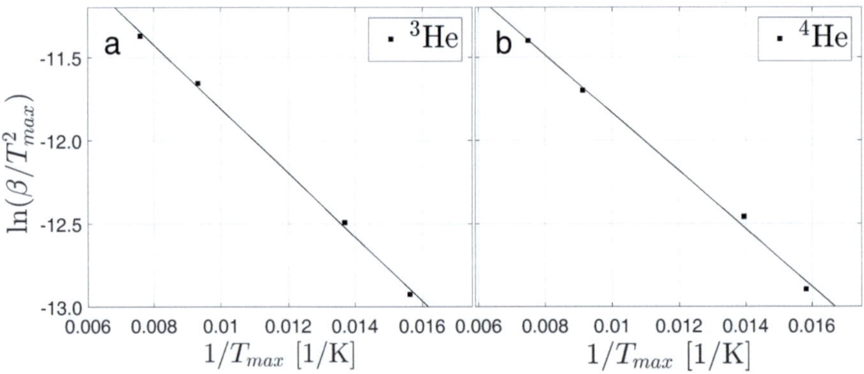

Figure 5.28: Kissinger plots for the determination of the desorption energy of ^{3}He (a) and ^{4}He (b) adsorbed in Br-MFU-4. β is the heating rate and T_{max} is the temperature of the maximum desorption rate.

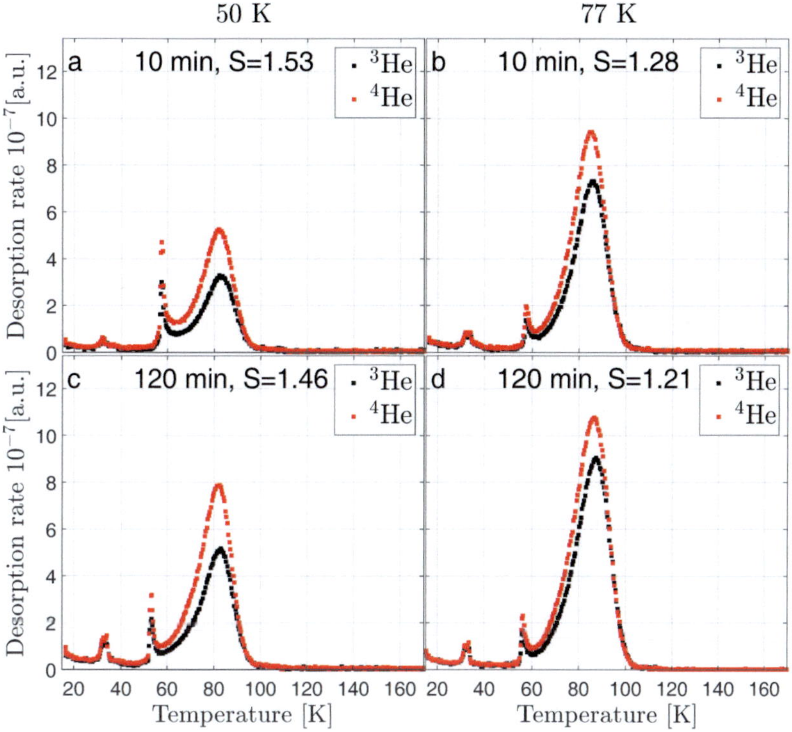

Figure 5.29: An equimolar mixture has been exposed to Br-MFU-4 at 50 K for 10 min (18 mbar), 120 min (22 mbar) and at 77 K for 10 min (19 mbar), 120 min (20 mbar). The respective selectivities are given in the plots. (heating rate: $0.1\,\mathrm{K\,s^{-1}}$)

Exposure to an equimolar isotope atmosphere has been performed at an exposure temperature of 50 K and 77 K for exposure times of 10 min and 120 min (figure 5.29). The maximum at low temperatures becomes very small and an additional very sharp peak rises in all spectra between 50 K and 60 K. The selectivities calculated from the total desorbed amount (desorption temperature range between 15 K and 300 K) are given in the diagrams.

5.3.2 Discussion

The Cl-MFU-4 structure has been described in chapter 4.2. Its small aperture (2.52 Å) between small and large pores is formed by four chlorine atoms. This aperture acts as a gate for gas molecules. For example, N_2 molecules cannot enter the pore structure, whereas the smaller molecule H_2 can penetrate into the pores. [18] Further investigations by Teufel *et al.* show that at very low exposure temperatures (30 K) even H_2 cannot enter the pores. [102] Phonon dispersion calculations revealed that the vibration of the chlorine atoms define the aperture size. These vibrations depend on temperature and, therefore, also the penetration of H_2 into the material is only observed above a certain temperature. Penetration into the material due to temperature dependent aperture size is called gate opening effect. [102]
Furthermore, the adsorption strength of gas molecules or atoms depends on the curvature of the pore wall (see chapter 2.1). This has been shown for H_2 by determining the heat of adsorption from isotherm measurements in MOFs with different pore sizes. [105] For MFU-4 possessing a bimodal pore size distribution with a large and a small pore, at low temperatures gas will be adsorbed in both pores, whereas with increasing temperature firstly less gas will be adsorbed in the large pore with the lower adsorption potential.
The combination of the gate opening effect and the bimodal pore size distribution will lead to an interesting adsorption and desorption behavior, depending on both exposure and desorption temperature.

Helium isotope adsorption in Cl-MFU-4

The highest amount of helium is adsorbed in Cl-MFU-4 already at 30 K exposure temperature, whereas at this temperature no adsorption occurs for H_2 [106]. At 30 K exposure temperature helium is adsorbed in both pores. With increasing desorption temperature more helium atoms are released from the surface of the large pore, which results in a higher internal pressure against the gate (aperture). This leads together with the increasing temperature during the TDS measurement to an opening of the gate and the desorbed helium atoms can diffuse out of the material. A sudden release may cause the sharp peaks at the beginning of the desorption spectra around 33 K. For different exposure times at 30 K, these sharp peaks occur at the same desorption temperature (see figure 5.19a), indicating the opening of the gate.

Increasing the exposure temperatures from 30 K to 60 K leads to a shift of the maximum to higher desorption temperatures (see figure 5.20a). At higher exposure temperatures the aperture size is larger and, therefore, the gas atoms can penetrate deeper. Additionally, at higher exposure temperatures the helium atoms are mainly bound at the pore walls of the small pores, which have higher adsorption potentials than the large pores (see chapter 2.1) and stronger bound atoms are released at higher temperatures. Due to adsorption only in the small pores, the smaller surface area of the small pores leads to a decreasing adsorbed amount of helium.
The whole amount of helium desorbes below 70 K (see figure 5.20a). Therefore, 60 K exposure temperature is the highest temperature at which helium atoms are still adsorbed at the wall of the small pore. The exposure to 70 K and higher temperatures shows a similar maximum (see figure 5.20b) occurring at slightly lower desorption temperatures than the maximum after exposure at 60 K. This behavior can be explained by fully open gates, in combination with the non-existent adsorption to the pore walls above 70 K, which can be substantiated by the following observations. The adsorbed amount at 77 K exposure temperature is constant for different exposure times, indicating that the diffusion into the material is time independent (see figure 5.19b). Evacuating the non-adsorbed gas at 77 K exposure temperature, leads to a complete removal of the desorption maximum (see figure 5.21b). In contrast, after evacuation at 40 K, the maximum is comparable with the maximum after evacuation at base temperature, which shows that no adsorption of helium atoms occurs at 77 K and the atoms can be pumped out easily through the open gate. During the exposure time at exposure temperatures above 70 K, the gas is randomly distributed in the pores and no adsorption takes place. The observed maxima in the TDS spectra show the amount of gas adsorbed during the cooling procedure. This adsorption of the randomly distributed gas in the pores starts when the temperature reaches values below 70 K. The cooling procedure below 70 K is independent on the exposure time or temperature applied before. Therefore, the adsorbed amounts and the maximum temperature are similar for the exposure temperatures above 70 K. At 298 K the gas density is significantly lower than at 90 K, leading to a smaller adsorbed amount.
Combining the obtained results with the dimensions of the aperture (2.52 Å) and the kinetic diameter of helium (2.60 Å), leads to the conclusion that the gate opening effect, which has been observed by Teufel *et al.* for hydrogen isotopes does not occur for helium. [102] Hydrogen isotopes as guest molecules could not penetrate into the material at 30 K exposure temperature. [106] The

first desorption signal occurred for the material exposed at 40 K and the maximum adsorption was reached for exposing the material at 60 K. The current investigations with helium show, that the adsorbed amount decreases for higher exposure temperatures. At zero temperature the aperture has a diameter of 2.52 Å. For non-zero temperatures, the atoms at the aperture vibrate, leading to an effective aperture that gets larger. Due to their smaller kinetic diameter, helium atoms (2.60 Å) can enter the pores at lower exposure temperatures than hydrogen (2.89 Å) and the gates of Cl-MFU-4 are open for the small helium atoms already at the lowest investigated exposure temperature of 20 K.

Helium isotope adsorption in Br-MFU-4

The observation that helium penetrates into Cl-MFU-4 already at 30 K gave the motivation to investigate helium adsorption in Br-MFU-4, which contains as gate atoms bromine instead of chlorine. By this substitution, the aperture is reduced to 2.33 Å. In the TDS spectra of Br-MFU-4 two maxima arise, one at lower temperature around 35 K and the other one at higher temperature around 106 K (table 5.6 and figure 5.24). The decrease of the desorption maximum with longer exposure times, which has been already measured for IFP-3 occurs here as well for the 35 K maximum (see figure 5.25). The bromine atoms in both materials restrict the aperture diameter. Therefore, the 35 K maximum can probably be attributed to the interaction between the bromine gate and the helium isotopes, whereas the high temperature signal arises from gas penetrating inside the pores. At 30 K exposure temperature no maximum arises around 100 K, indicating that helium atoms cannot penetrate into the material (figure 5.26a,c). After exposure at 40 K a second maximum appears in the TDS spectrum at 94 K. For exposure at 50 K this maximum first increases and then decreases for higher exposure temperatures. The influence of the gas penetration on the exposure temperature has also been observed by Teufel *et al.* for hydrogen isotopes adsorbed in the structure of Cl-MFU-4 and has been denoted as gate opening effect. [102]

Desorption energy of helium isotopes adsorbed in Br-MFU-4

In this work the heating rate variation measurements with TDS have been used to investigate the desorption energy of ^{3}He and ^{4}He in Br-MFU-4 by evaluating the Kissinger plots presented in figure 5.28. From their slope the desorption energies of 1.6 kJ mol^{-1} for ^{3}He and 1.5 kJ mol^{-1} for ^{4}He have been determined

(see chapter 2.3). For helium adsorption less can be found in literature. Martins *et al.* have measured adsorption isotherms of ^{4}He in activated carbons and determined an adsorption energy between 2.2 kJ mol^{-1} and 2.5 kJ mol^{-1}. [107] In Na-Y zeolite an adsorption energy of 1.4 kJ mol^{-1} and 1.6 kJ mol^{-1} for ^{3}He and ^{4}He, respectively, has been determined by the heat pulse technique. [108] The obtained values for Br-MFU-4 are similar to the values found in literature for activated carbon and Na-Y zeolite.
The comparable desorption temperature for the maximum of IFP-4 and Br-MFU-4 (around 100 K) would imply that the desorption energies of both materials are similar for helium isotopes. However, 5.1 kJ mol^{-1} and 1.5 kJ mol^{-1} have been obtained here with the Kissinger method for IFP-4 (see chapter 5.2.2) and Br-MFU-4, respectively. Helium atoms are still adsorbed in IFP-4 if the evacuation of the non-adsorbed gas is done at 77 K, whereas all atoms are removed in Br-MFU-4 after evacuation at 77 K. This is in accordance with the higher value for the desorption energy of IFP-4.

Separation of helium isotope in Cl-MFU-4 and Br-MFU-4

Cl-MFU-4

Measurements with equimolar mixtures of helium isotopes have shown that the highest selectivity of $S = 2.3$ is obtained when applying 17 mbar at 30 K, for 10 min. When exposing at 30 K, the selectivity is almost independent of applied pressure and exposure time (see table 5.5). For higher exposure temperatures, the aperture widens and the selectivity drops below $S = 2$.

Br-MFU-4

For hydrogen isotopes exposed to Cl-MFU-4, the highest value for the selectivity of $S = 7.5$ has been obtained at 60 K exposure temperature. At this exposure temperature also the highest amount of hydrogen isotopes has been adsorbed. [102] Here, the highest adsorption of helium isotopes occurs at an exposure temperature of 50 K (see figure 5.26) and, therefore, this temperature and the liquid nitrogen temperature (77 K) have been chosen as exposure temperatures for the investigation of selective adsorption. The selectivity values are below $S = 2$ and decrease slightly for longer exposure times (figure 5.29). The TDS spectra after applying isotope mixture differ significantly from the measurements with pure gas: 1. the amount of adsorbed gas at 77 K is higher than at 50 K, 2. the high temperature desorption maximum is at lower desorption temperatures and 3. the low temperature desorption maximum almost disappeared. Due to the low selectivity values, no further measurements with the precious ^{3}He have been performed for the clarification of the different spectra if mixtures are applied.

Summary

For the separation of helium isotopes with Quantum Sieving, two materials of the MFU-4 series have been chosen. Cl-MFU-4 (2.52 Å) appeared promising as it has a high selectivity for hydrogen isotopes [102]. Br-MFU-4 (2.33 Å), with an even smaller aperture diameter, appeared promising for separation of the smaller helium isotopes (2.60 Å). The different aperture sizes have a direct influence on the adsorption behavior. Whereas for Cl-MFU-4 the highest adsorbed amount occurred at the lowest exposure temperature (30 K), for Br-MFU-4 at that temperature no helium isotopes could penetrate into the material, but with higher exposure temperature, the flexible apertures become larger and allow adsorption.

The highest selectivity of $S = 2.3$ has been obtained with Cl-MFU-4 after exposing a helium isotope equimolar mixture at 30 K. However, the high selectivity result obtained for hydrogen isotopes of $S = 7.5$ [102] is by far not reached. This can be understood by the smaller mass difference between ^{3}He and ^{4}He, compared to H_2 and D_2, which results in a smaller difference in zero-point energy. Therefore, the difference in diffusivity of ^{3}He and ^{4}He is smaller and the separation by Quantum Sieving is lower.

6 Hydrogen isotope adsorption and separation in MOFs with open metal sites

6.1 Cu(I)-MFU-4*l*

6.1.1 Thermal desorption spectroscopy

Results

In preparation for this series of measurements, Cu(I)-MFU-4*l* has been activated in the TDS sample chamber at 453 K for one hour. This material has been synthesized by post-synthetical exchange of Zn(II)-Cl units of MFU-4*l* with Cu(I) (see chapter 4.3). Measurements with the pristine material MFU-4*l* have been performed in order to identify the changes in the TDS spectrum occurring due to adsorption to the Cu(I)-open metal site. This material has been activated at 423 K for two hours. All TDS spectra have been recorded with a heating rate of $0.1\,\mathrm{K\,s^{-1}}$ if it is not explicitly mentioned.
Cu(I)-MFU-4*l* has been exposed to 15 mbar pure H_2 or pure D_2 at room temperature and after cooling to 20 K under gas atmosphere, the non-adsorbed gas has been evacuated. The thermal desorption spectrum shows two maxima at low temperatures (30 K, 50 K), which are similar to the maxima of MFU-4*l* (see figure 6.2). In the temperature region above 100 K only Cu(I)-MFU-4*l* exhibits additional maxima at 173 K for H_2 and 181 K for D_2, resulting from strongly bound hydrogen isotopes at the open metal sites.

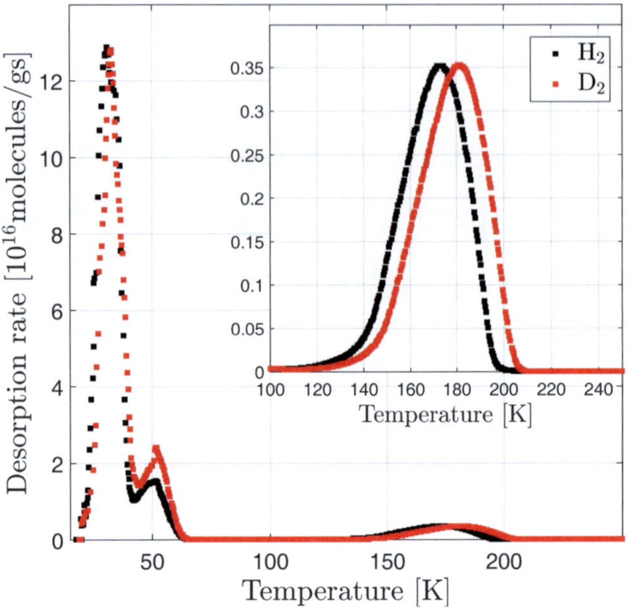

Figure 6.1: TDS spectra of Cu(I)-MFU-4*l* exposed to 15 mbar H_2 or D_2 at room temperature. The inset augments the desorption rate at high temperatures. (heating rate: $0.1\,K\,s^{-1}$)

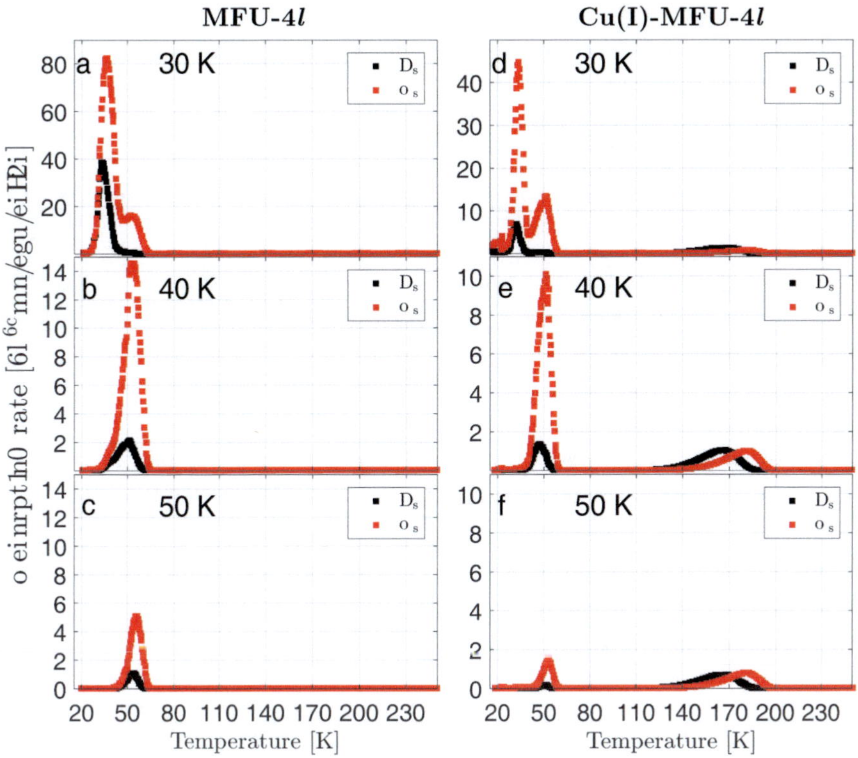

Figure 6.2: Comparison between MFU-4*l* and Cu(I)-MFU-4*l*. An equimolar mixture of 10 mbar has been loaded for 10 min to the samples at 30 K, 40 K and 50 K. Please note the varying y-axes scalings. (heating rate: $0.1\,\mathrm{K\,s^{-1}}$)

Figure 6.2 shows a comparison of the TDS spectra for MFU-4*l* and Cu(I)-MFU-4*l* after applying identical conditions (10 mbar, 10 min) during exposure to an equimolar isotope mixture at different exposure temperatures (30 K, 40 K, 50 K). The non-adsorbed gas has been evacuated at exposure temperature and, therefore, the adsorbed amount decreases for higher exposure temperatures.
The following results are focused on the Cu(I)-open metal site.
For the determination of the desorption energy of both isotopes from the Cu(I)-open metal site by TDS, the following measurements have been performed. The sample has been loaded at 100 K exposure temperature with an equimolar mix-

ture of 10 mbar for 10 min. Different heating rates of $0.1\,\mathrm{K\,s^{-1}}$, $0.05\,\mathrm{K\,s^{-1}}$ and $0.01\,\mathrm{K\,s^{-1}}$ have been applied for the desorption process and the TDS spectra are presented in figure 6.3a. In figure 6.3b, the data have been plotted with the unit molecules per gram kelvin for better visibility.

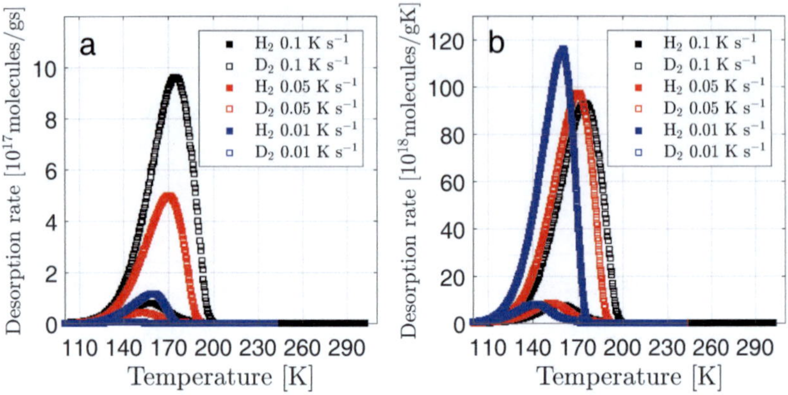

Figure 6.3: At 100 K exposure temperature the sample has been loaded with 10 mbar isotope mixture for 10 min. The desorption spectrum has been recorded with different heating rates: $0.1\,\mathrm{K\,s^{-1}}$, $0.05\,\mathrm{K\,s^{-1}}$ and $0.01\,\mathrm{K\,s^{-1}}$. a) Desorption rate in molecules per gram seconds. b) Desorption rate in molecules per gram kelvin.

The desorption maximum is shifted to lower temperatures for slower heating rates, which indicates that the release of the isotopes is thermally activated. Application of the Kissinger method allows to determine the desorption energy of hydrogen isotopes adsorbed at the open metal site. This method can be used according to equation 2.20 and equation 2.21 in two ways, which are presented in figure 6.4. T_{max} is the temperature at the maximal desorption rate, β is the heating rate and N_p is the value of the desorption rate at T_{max}. As described in chapter 2.3, the desorption energy can be extracted from the slope of the linear fit presented in figure 6.4.

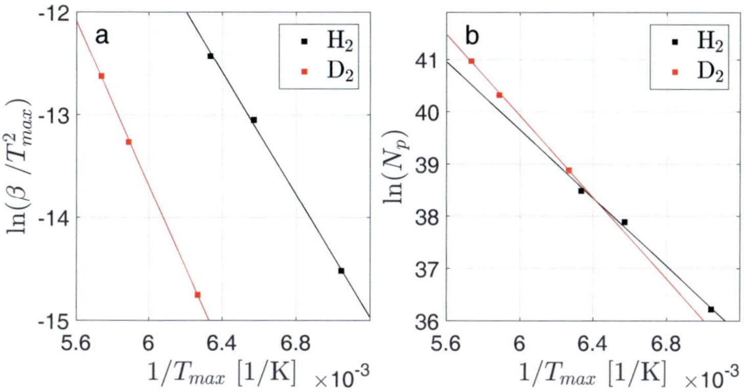

Figure 6.4: Determination of the desorption energy with the heating rate variation method by a) using the desorption rate maximum position on the x-axis (T_{max}) combined with the heating rate (β), b) using the desorption rate maximum position on the y-axis (N_p).

The obtained desorption energies are summed up in table 6.1.

	E_{des} in kJ mol^{-1} according to Eq. 2.20 (figure 6.4a)	E_{des} in kJ mol^{-1} according to Eq. 2.21 (figure 6.4b)
H_2	24.8	27.1
D_2	33.2	32.5

Table 6.1: Desorption energy of H_2 and D_2 adsorbed at the open metal site of Cu(I)-MFU-4*l*.

Figure 6.5 shows the TDS spectra of Cu(I)-MFU-4*l* after it has been exposed to an equimolar hydrogen isotope mixture of 10 mbar for 10 min at different exposure temperatures. At low exposure temperatures the H_2 signal dominates the desorption spectrum, indicating that more H_2 was adsorbed. For 40 K and 50 K exposure temperature the desorption maxima of both isotopes are nearly equal showing that at these exposure temperatures almost the same amount of H_2 and D_2 is adsorbed. At higher exposure temperatures more D_2 than H_2 is adsorbed resulting in an increase of the D_2 desorption maximum.

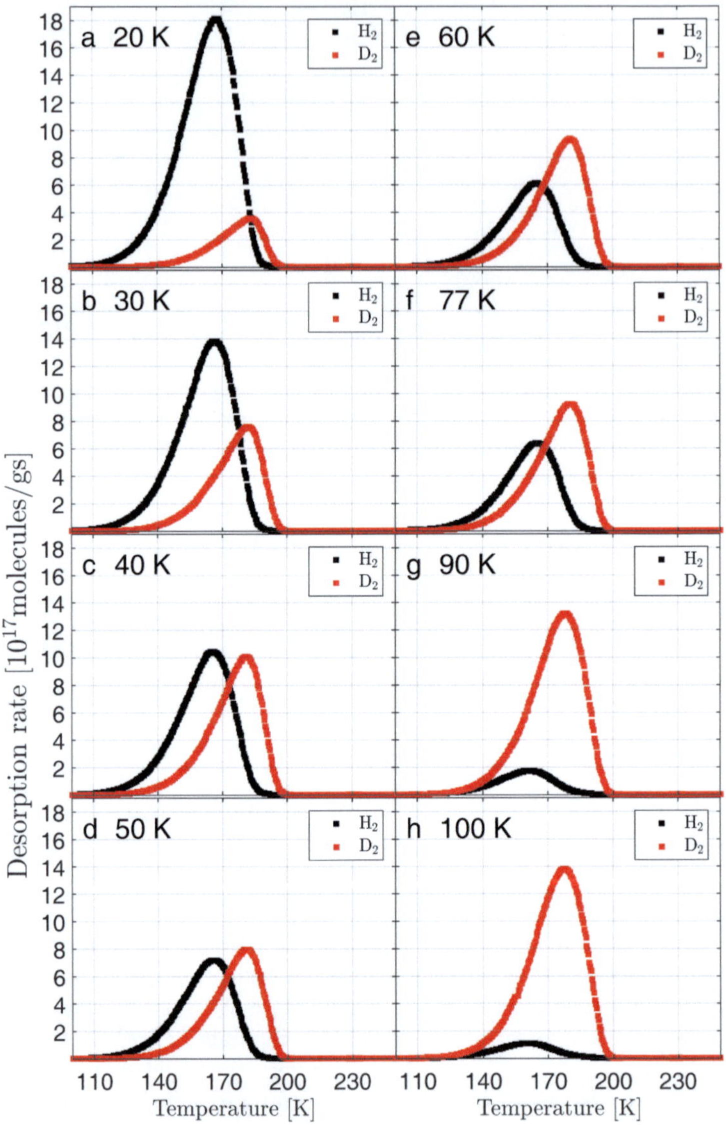

Figure 6.5: H_2 and D_2 desorption spectra after exposure to an equimolar mixture of 10 mbar for 10 min at different exposure temperatures. (heating rate: $0.1\,\mathrm{K\,s^{-1}}$)

The ratio of the released D_2 and H_2 amount directly gives the selectivity (see chapter 3.1.3). The selectivity as a function of the exposure temperature is shown on the left side of figure 6.6 considering the total desorbed gas, i.e. the whole desorption temperature range (from 15 K to 300 K). 20 K exposure temperature leads to a high gas release at low desorption temperatures, which is above the maximum range of the mass spectrometer and therefore no value can be given for the selectivity at 20 K considering all adsorption sites. The high maxima at 30 K and 50 K desorption temperature have the major impact for the selectivity of the material at exposure temperatures up to 40 K (see figure 6.2d,e). For exposure temperatures higher than 50 K only the strong binding open metal site is occupied by hydrogen isotopes. The selectivity resulting only from the adsorbed amount at the open metal sites (temperature range from 100 K to 300 K) is shown in figure 6.6 right. The adsorbed amount of H_2 exceeds the adsorbed amount of D_2 at low exposure temperatures and at higher exposure temperatures a preferential adsorption of D_2 occurs. For this reason the selectivity (D_2/H_2 ratio) is given in a logarithmic scale.

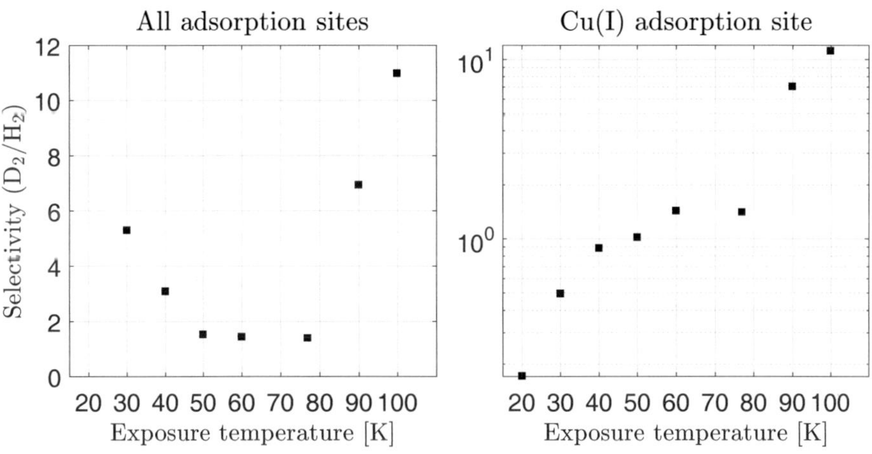

Figure 6.6: The selectivity is plotted in dependence of the exposure temperature. Left: All desorbed gas is considered for the calculation of the selectivity. Right: Only the desorbed gas above 100 K desorption temperature is considered for the calculation of the selectivity.

The high selectivity value at 100 K exposure temperature, motivated to investigate the time dependence of the selectivity. Figure 6.7a shows TDS spectra after the sample has been exposed to an equimolar mixture of 10 mbar for different exposure times. The selectivity obtained from the TDS spectra shown in figure 6.7a is presented in figure 6.7b and shows a linear increase within the investigated exposure times.

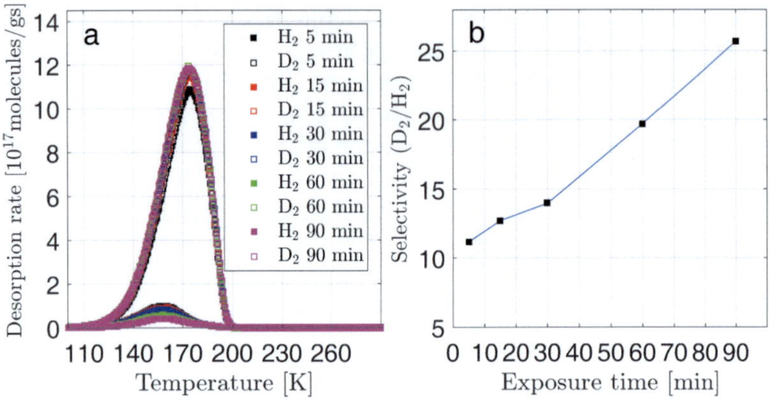

Figure 6.7: a) TDS spectra of the open metal site exposed to an equimolar mixture of 10 mbar for different exposure times at 100 K. b) Selectivity of the open metal site in dependence on exposure time. The blue line is a guide for the eye. (heating rate: $0.1\,K\,s^{-1}$)

The change from preferred H_2 adsorption to preferred D_2 adsorption with rising exposure temperature has never been observed before. Therefore, a new method, called 'stepwise isotope loading', was applied (see chapter 3.1.3, figure 3.4). In the first step, the sample is exposed only to gas of isotope 1. After 10 min exposure time, isotope 2 is added and the sample is then exposed to an isotope mixture. In figure 6.8 the high-temperature desorption maximum is shown for a stepwise loading with the two isotopes at different exposure temperatures: 30 K, 77 K and 100 K.

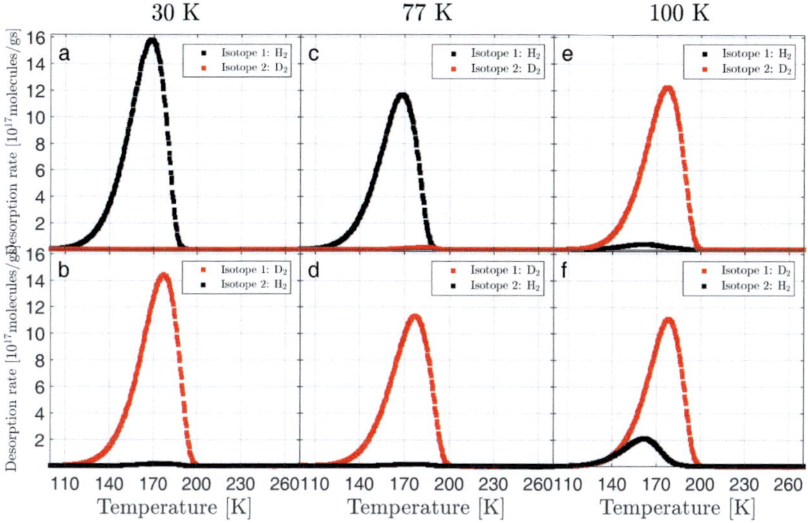

Figure 6.8: Investigation of the open metal site with the stepwise gas loading procedure. 5 mbar of the first isotope has been added for 10 min. Subsequently 5 mbar of the second isotope has been added for 10 min. (heating rate: $0.1\,K\,s^{-1}$)

The upper row shows the spectra after exposing firstly to H_2 and then adding D_2 to form a gas mixture. In the lower row the loading sequence is changed and the first exposure was to D_2 and than H_2 was added. At 30 K and 77 K exposure temperature the desorption spectrum consists almost solely of isotope 1. In contrast, at 100 K exposure temperature the D_2 signal is for both loading sequences the higher one. This high D_2 signal is especially astonishing for the case of the sample being firstly exposed to H_2. The difference in the heights of the H_2 signals in figures 6.8e,f is caused by the different ratio of the isotope mixtures applied in the second loading step. During the stepwise loading procedure, experimental conditions (see section 3.1), lead to a ratio of 1:2.5 for isotope 1:isotope 2.

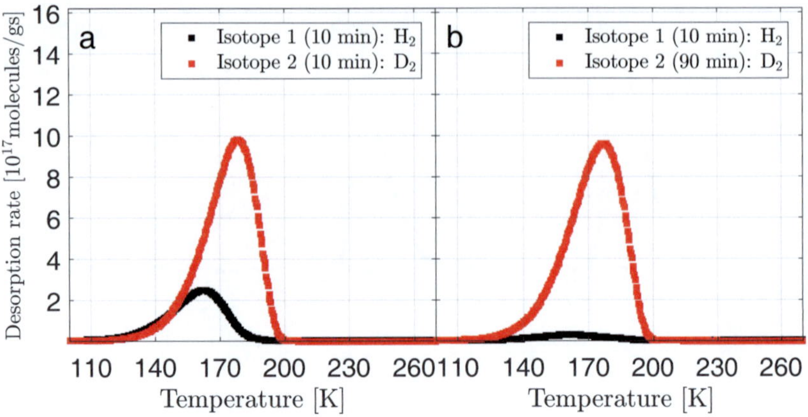

Figure 6.9: Stepwise gas loading procedure at 90 K with different exposure times for the second step: 10 min (a) and 90 min (b). The applied pressure is 5 mbar in each step. (heating rate: $0.1\,K\,s^{-1}$)

Additionally, the stepwise loading has been performed with an exposure time of 10 min for the first step and 90 min for the second step. Figure 6.9 shows the TDS spectra after adding 5 mbar of H_2 as isotope 1 for 10 min and 5 mbar of D_2 as isotope 2 for 10 min (a) and 90 min (b) at 90 K. It is visible that for longer exposure times in the second step, where the sample is exposed to an isotope mixture, the H_2 signal is strongly reduced.

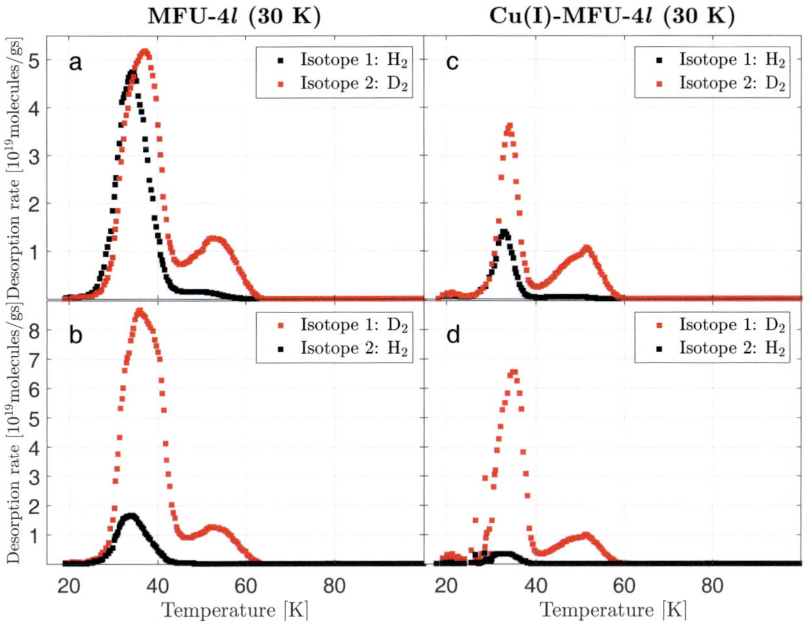

Figure 6.10: TDS spectra of the stepwise loading procedure for the low temperature adsorption sites in MFU-4*l* and Cu(I)-MFU-4*l*. The samples have been exposed at 30 K to 5 mbar of each isotope with an exposure time of 10 min for each step. (heating rate: $0.1\,\mathrm{K\,s^{-1}}$)

The desorption spectra in the low temperature regime have also been measured with the stepwise loading procedure. Figure 6.10 shows the results at 30 K for MFU-4*l* on the left side and Cu(I)-MFU-4*l* on the right side and figure 6.11 shows the results at 50 K exposure temperature, respectively. These investigations reveal that D_2 always exceeds the H_2 signal both when it is added firstly or secondly.

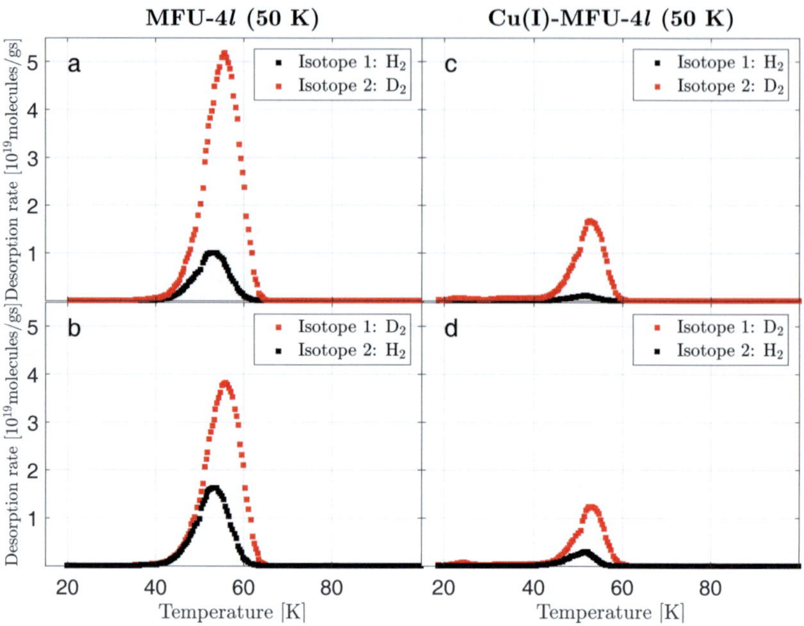

Figure 6.11: TDS spectra of the stepwise loading procedure for the low temperature adsorption sites in MFU-4*l* and Cu(I)-MFU-4*l*. The samples have been exposed at 50 K to 5 mbar of each isotope with an exposure time of 10 min for each step. (heating rate: $0.1\,\mathrm{K\,s^{-1}}$)

The desorption spectrum of the open metal site loaded at 40 K exposure temperature with a 50% H_2 and 50% D_2 (equimolar) mixture shows an almost equal value for the maximum desorption rates of the isotopes (figure 6.12a). Loading with a 30% H_2 and 70% D_2 mixture leads to a lower desorption maximum for H_2 compared to D_2 (figure 6.12b).

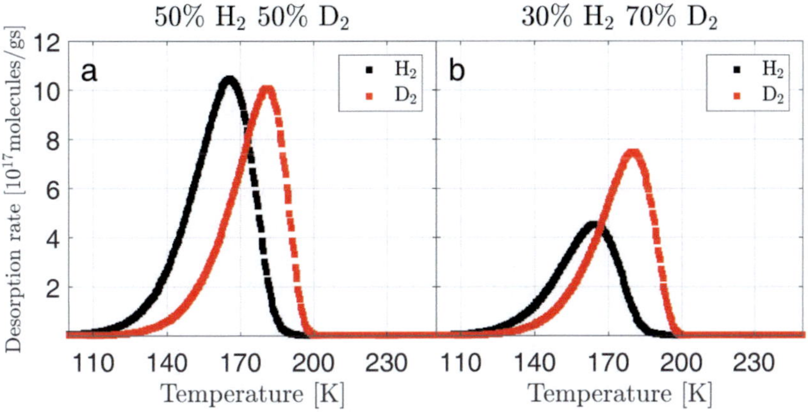

Figure 6.12: a) 10 mbar of an equimolar mixture has been loaded at 40 K for 10 min. b) 10 mbar of a mixture containing 30% of H_2 and 70% of D_2 has been loaded at 40 K for 10 min. (heating rate: $0.1\,K\,s^{-1}$)

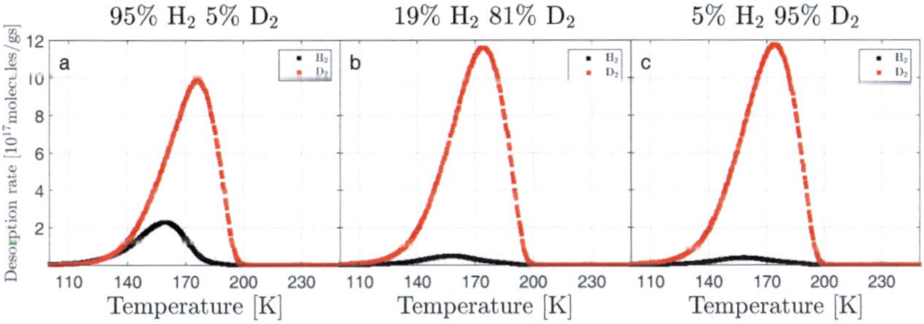

Figure 6.13: Non-equimolar mixtures of 10 mbar have been applied at 100 K for 10 min. A concentration of 95% H_2 and 5% D_2 has been chosen as starting point. The subsequent concentrations follow the foregoing separation factor. (heating rate: $0.1\,K\,s^{-1}$)

To study the extraction of D_2 from an initial low concentration of a H_2:D_2 mixture at the open metal site in more detail, the sample was exposed at 100 K to an isotope mixture containing only 5% D_2 and 95% H_2. Figure 6.13a shows the resulting desorption spectrum of H_2 and D_2 with a concentration of 81% D_2 to 19% H_2. This concentration was applied in a second exposure step and the resulting desorption spectrum is shown in figure 6.13b giving now a concentration of 93% D_2 to 7% H_2. In a third process cycle gas with this concentration was exposed to the sample at 100 K resulting in a concentration of the desorbed gas of 95% D_2 to 5% H_2 (figure 6.13c).

Discussion

The TDS results of Cu(I)-MFU-4*l* after exposure to hydrogen isotopes and isotope mixtures will be discussed and correlated to the structure of this novel material by comparison with the pristine MFU-4*l*.

Adsorption sites of MFU-4*l*

The structure of pristine MFU-4*l* is well known (see chapter 4.3) and it has already been investigated by TDS in previous studies. [95,106,109] Calculations with the method of density functional theory have been performed revealing the most favorable adsorption sites for hydrogen isotopes in MFU-4*l*. [11]

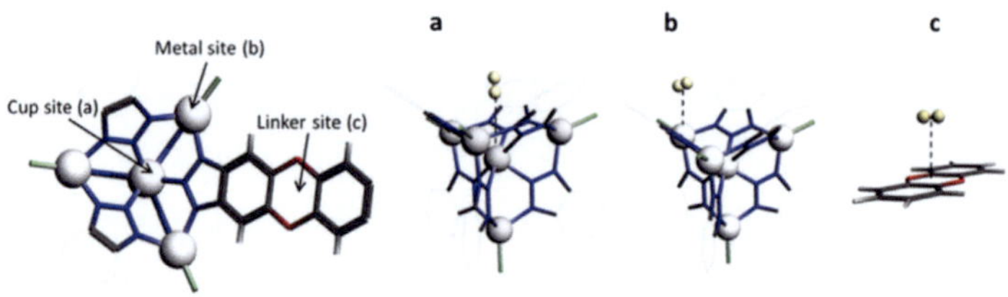

Figure 6.14: Adsorption sites of MFU-4*l*. a) is the cup site with a linear orientation of the H_2 with respect to the central zinc atom, b) is the metal site with a T-shaped orientation of H_2 with respect to the Zn-Cl unit and c) is the linker site, where H_2 is oriented in-plane with the linker site along the oxygen-oxygen axis. [11]

Three sites, named cup site ($10.5\,\mathrm{kJ\,mol^{-1}}$), metal site ($8.7\,\mathrm{kJ\,mol^{-1}}$) and linker site ($5.5\,\mathrm{kJ\,mol^{-1}}$) with descending binding strength have been found and are shown in figure 6.14. [11]

Comparison between MFU-4l and Cu(I)-MFU-4l

Cu(I)-MFU-4*l* and the pristine MFU-4*l* have basically the same framework structure. [19] For the preparation of Cu(I)-MFU-4*l* the tetrahedral coordinated zinc centers of MFU-4*l* have been post-synthetically replaced by copper. Thus, this exchange of approximately 30% of the Zn(II)-Cl units directly reduces the number of the metal (zinc) site. [110] Identical experimental conditions have been applied to both materials in order to reveal how the Cu(I)-open metal site influences the pristine adsorption sites.
In the low temperature region up to 100 K desorption temperature, two maxima rise in the spectra of Cu(I)-MFU-4*l* and MFU-4*l* at the same desorption temperatures. Hence, the same adsorption sites exist in both materials (see figure 6.2). In pristine MFU-4*l*, above 70 K, no hydrogen isotope desorption occurs, which is in accordance with former investigations. [95, 106] The strongest adsorption site of MFU-4*l* is the cup site, which is ascribed to the maximum at 50 K. Isotopes desorbing from the linker and zinc-metal site, appear in the maximum at 30 K. A comparison of the maximum height resulting from these sites (see figure 6.2a,d) shows that MFU-4*l* reaches approximately 80 molecules/gs, whereas Cu(I)-MFU-4*l* has a lower value of 50 molecules/gs. The linker site is the same in both materials, therefore, the lower signal for Cu(I)-MFU-4*l* directly reveals the reduced number of zinc-metal atoms due to the post-synthetic exchange with copper.
The main difference in the spectra of MFU-4*l* and Cu(I)-MFU-4*l* occurs at desorption temperatures above 100 K. Only for Cu(I)-MFU-4*l* a desorption maximum appears around 175 K. The high desorption temperature reveals that the isotopes are bound to a very strong adsorption site, the Cu(I)-open metal site.

Desorption energy of the open metal site of Cu(I)-MFU-4l

In this work, thermal desorption spectra recorded with different heating rates are used to determine the desorption energy with the Kissinger method (see figures 6.3, 6.4 and chapter 2.3). This method can be applied under the assumption that adsorption and desorption are reversible processes and the desorption energy is independent of coverage or temperature. The desorption energies for both hydrogen isotopes have been determined according to equation 2.20 and equation 2.21 and are presented in table 6.2.

	E_{des} [kJ mol^{-1}] TDS (figure 6.4a)	E_{des} [kJ mol^{-1}] TDS (figure 6.4b)	E_{des} [kJ mol^{-1}] isotherm measurements	E_{des} [kJ mol^{-1}] DFT calculations
H_2	24.8	27.1	32.5 [19]	25 [19]
D_2	33.2	32.5	35.0 [111]	–

Table 6.2: Desorption energy of H_2 and D_2 adsorbed at the open metal site of Cu(I)-MFU-4*l*.

The measurements performed in this work with TDS are in the same range as values known from literature (see table 6.2). Also all measurements show the same tendency, as the heavier isotope with the lower-zero point energy has a higher desorption energy. This is as expected, from TDS spectra, where the D_2 maximum appears at higher desorption temperatures. Cu(I)-MFU-4*l* shows up to now the highest known value for hydrogen isotopes adsorbed in MOFs.

Hydrogen isotope adsorption on the open metal site of Cu(I)-MFU-4*l*

Investigations of exposure time dependence of the adsorption at the open metal site are shown in figure 6.7a. The large diameters of 11.97 Å and 18.56 Å and the aperture of 9.13 Å, do not hinder the diffusion of hydrogen isotopes into the material (see chapter 4.3). Therefore, only small time dependence has been observed in the thermal desorption spectra at the open metal site.
When pure H_2 or pure D_2 isotopes are applied to Cu(I)-MFU-4*l*, shape and height of the resulting desorption spectra of the open metal sites are almost identical for both gases (figure 6.1). Though, due to the higher desorption energy of D_2, the desorption maximum is reached at higher temperatures, compared to H_2.
H_2 or D_2 can also be applied sequentially, using a method developed during this work called 'stepwise loading procedure'. When D_2 is applied firstly, the subsequently applied H_2 can at no temperature replace the already adsorbed D_2 (see figure 6.8b,d,f). However, when H_2 is applied firstly, the subsequently applied D_2 can in principle replace the adsorbed H_2 due to its lower zero point energy, but at low exposure temperatures up to 77 K, the thermal energy for this exchange is not provided and H_2 remains adsorbed (figure 6.8a,c). For increasing temperature, the exchange in favor of the heavier isotope with the lower zero-point energy becomes possible (figure 6.8e), which is even further

increased if longer exposure times are applied for the second exposure step (compare figure 6.9a,b).
When mixtures of H_2 or D_2 are applied, the resulting spectra give further insight into the adsorption and desorption behavior. For higher exposure temperatures (>77 K), the TDS spectrum confirms, that at the open metal site the isotope with the lower zero point energy is adsorbed preferentially. The exchange observed with the stepwise loading procedure here also occurs, as longer exposure times lead to a higher amount of adsorbed D_2 along with a lower amount of adsorbed H_2. This is quantified by the selectivity, which increases from $S = 10.8$ to $S = 25.2$ when the exposure time is increased from 5 min to 90 min (figure 6.7b). The selectivities at high temperatures are in general larger than one (S > 1), which here is emphasized once more, as the situation in case of Cu(I)-MFU-4*l* is drastically different for low temperatures (20 K and 30 K). There, the relative difference in diffusion rates of D_2 and H_2 is largest. When the sample is exposed to the isotope mixture, the H_2 isotopes arrive at the strong adsorption sites much faster than the D_2 isotopes and occupy the strong adsorption sites before the slower diffusing D_2 isotopes arrive. As observed with the stepwise loading procedure, at these exposure temperatures the thermal energy for the exchange of H_2 by D_2 is not provided and the strong adsorption site remains occupied by H_2. The resulting selectivities, when exposing at low temperatures, are smaller than one (S < 1) despite the preferential adsorption of D_2. The transition between the high and the low temperature regime is found between 40 K and 77 K, where the larger diffusion rate of H_2 is counterbalanced by the lower zero point energy of D_2. There the selectivity is close to one (S ≈ 1) and the open metal site is equally occupid by both isotopes, which reflects the applied equimolar mixture. When the isotope ratio of the applied mixture is changed, the desorbed amount changes likewise (see figure 6.12).

Separation of hydrogen isotopes with the open metal site of Cu(I)-MFU-4*l*

In principle, adsorption in MOFs works better with lower temperature, which is why all presented investigations were performed far below room temperature. For the total amount of adsorbed gas, this also holds in Cu(I)-MFU-4*l* (figure 6.15b), as at higher temperatures the particles have higher energies allowing them to desorb more easily. However, as shown in figure 6.15a, in Cu(I)-MFU-4*l* the adsorbed amount of only D_2 as well as the selectivity increase with increasing exposure temperatures. This behavior is explained with the aforementioned exchange process, which happens predominantly at higher temperatures and makes the material even more suitable for isotope separation at high temperatures. It is a unique selling point of Cu(I)-MFU-4*l*, not shown in other materials. For comparison, the study of Oh *et al.* on CPO-27-Co can be mentioned, where the selectivity increases for higher exposure temperatures but the total amount of adsorbed D_2 is decreased. [44] Thus, in CPO-27-Co, unlike in Cu(I)-MFU-4*l*, a tradeoff between high selectivity and high adsorbed amount of D_2 must be made.

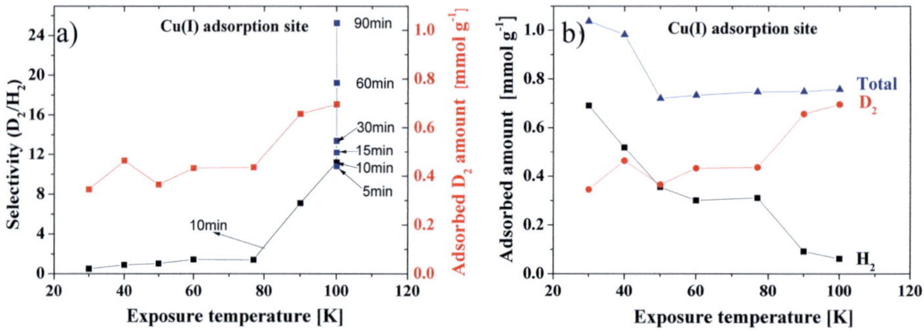

Figure 6.15: a) On the left axis, the selectivity of Cu(I)-MFU-4*l* is shown for different exposure temperatures after 10 min exposure time (black) and at 100 K for different exposure times (blue). The right axis shows the adsorbed amount of D_2 in dependence on the exposure temperature (red). b) Amount of H_2 (black), D_2 (red) and their sum (blue) adsorbed at the copper open metal site for different exposure temperatures, after applying 10 mbar equimolar hydrogen isotope mixture for 10 min to the sample.

Higher exposure temperatures than 100 K have not been investigated, as evacuation at exposure temperature would lead to a partial removal of molecules adsorbed at the Cu(I)-open metal site, which start to desorbe slightly above 100 K in the TDS spectrum. Hence, for obtaining the high selectivity at the open metal site cooling below 100 K is not necessary, and for a potential application the expensive liquid He as cooling agent can be replaced by liquid N_2. The strong Cu(I)-open metal site could be used for industrial isotope separation in the following way:

1. Load a hydrogen isotope mixture at 100 K, which contains a small amount of D_2.
2. Pump out the not adsorbed gas at 100 K.
3. Start the desorption process by applying a temperature ramp up to 200 K.
4. Store the released gas amount in a reservoir.
5. Cool back to 100 K.
6. Connect the reservoir to the sample chamber, to load the isotope mixture again and repeat the process.

The setup of the TDS device is up to now not suitable for storing the desorbed amount. Therefore, for following cycles a mixture containing the measured isotope concentration obtained from the TDS spectrum of the previous cycle has been exposed. After the first cycle, where only 5% D_2 have been exposed for 10 min exposure time, the concentration of D_2 in the desorbed gas amount increased to 81% (figure 6.13). A concentration of 95% D_2 has been received after only three cycles. Increasing the exposure time has shown an increase for the selectivity from $S = 11.0$ to $S = 25.2$ for 10 min and 90 min exposure time (figure 6.7b), respectively. Thus, longer exposure times can reduce the required number of cycles for the extraction of D_2 from an initially low concentration.

Summary

The copper open metal site of Cu(I)-MFU-4*l* binds both H_2 and D_2 very strongly. Desorption energies of 24.8 kJ mol^{-1} for H_2 and of 33.2 kJ mol^{-1} for D_2 have been determined in this work, which are among the highest values for MOFs to our knowledge. Due to this very strong adsorption, the maxima in the TDS spectrum occur at very high temperatures, namely at 166 K for H_2 and at 181 K for D_2. The heavier isotope desorbs at higher temperatures, which is in accordance with its higher desorption energy due to the lower zero-point energy.

It has been observed for the first time, that the adsorbed amount from an equimolar isotope mixture depends on the difference in zero-point energies of the isotopes, combined with the diffusion to the adsorption sites and their strength. The faster diffusing, lighter isotope reaches the strong adsorption sites first and cannot be exchanged by the later arriving heavier isotope at low exposure temperatures of 20 K and 30 K. Introducing the isotopes one after

the other to the sample shows that up to 77 K the first added isotope remains adsorbed. For higher temperatures, D_2, with the lower zero-point energy, is always preferentially adsorbed. Therefore, at exposure temperatures above 77 K an exchange of the already adsorbed H_2 by D_2 occurs, whereas at low exposure temperatures no exchange is possible and the first arriving isotope from an isotope mixture remains adsorbed.
Due to the time dependent exchange process of H_2 by D_2 at the strong adsorption site, the selectivity of hydrogen isotopes increases linearly with time from 5 min to 90 min, reaching the highest selectivity of $S = 25.2$ (at 100 K). Such a high exposure temperature allows to replace the expensive liquid helium as coolant with liquid nitrogen, which is essential for potential industrial applications. This has been investigated by loading a mixture with an initially low concentration of D_2. Already after the first adsorption/desorption cycle, the concentration increased from 5% in the applied gas to 81% in the desorbed gas.

6.1.2 Inelastic neutron spectroscopy

The inelastic neutron spectroscopy measurements presented in this work have been performed at the VISION spectrometer of the Oak Ridge National laboratory. They were the first measurements ever performed with gas loading to a sample cell on this spectrometer. VISION is up to now the highest resolution broadband inelastic neutron scattering spectrometer in the world with an overall inelastic count rate of more than two orders of magnitude compared to similar spectrometers. [67] This spectrometer is a analogue of a Raman or infrared spectrometer with a broad energy range of -2 to 1000 meV.

Results

The goal of measuring the transitions of hydrogen isotopes adsorbed in Cu(I)-MFU-4*l* with inelastic neutron scattering (INS) is to study their rotational energy spectrum at different adsorption sites. First, measurements of the metal-organic framework in the aluminum sample holder have been performed without gas loading. This background signal has been subtracted for all presented figures. Figure 6.16 shows the INS spectra of hydrogen adsorbed in Cu(I)-MFU-4*l* at 5 K. The gas amounts are given relative to each other. Under low hydrogen loading a double peak at 4.9 meV and 6.5 meV and a broad maximum around 37 meV appear in the spectrum. The higher loading leads to a second double peak at 13.4 meV and 14.7 meV, additional peaks at 23.0 meV and 28.7 meV and an increased scattering intensity above 30 meV.

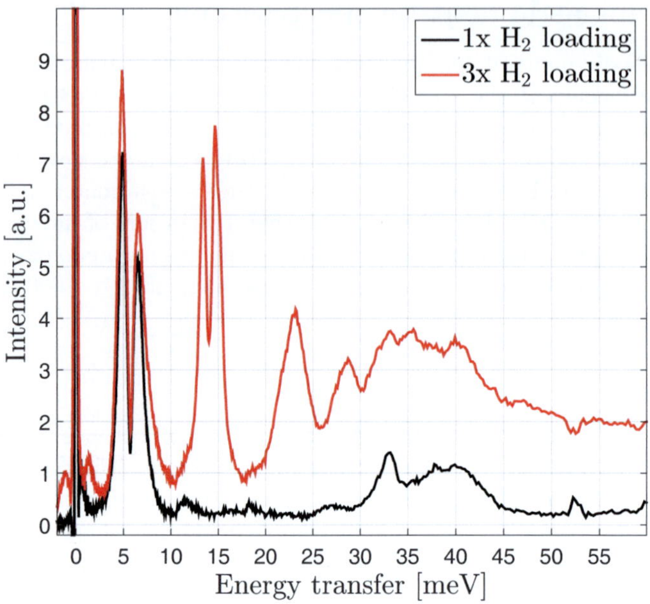

Figure 6.16: Inelastic neutron scattering of H_2 adsorbed to Cu(I)-MFU-4*l*. Two different amounts have been loaded to the sample, which are given in relation to each other. The spectra have been recorded at 5 K.

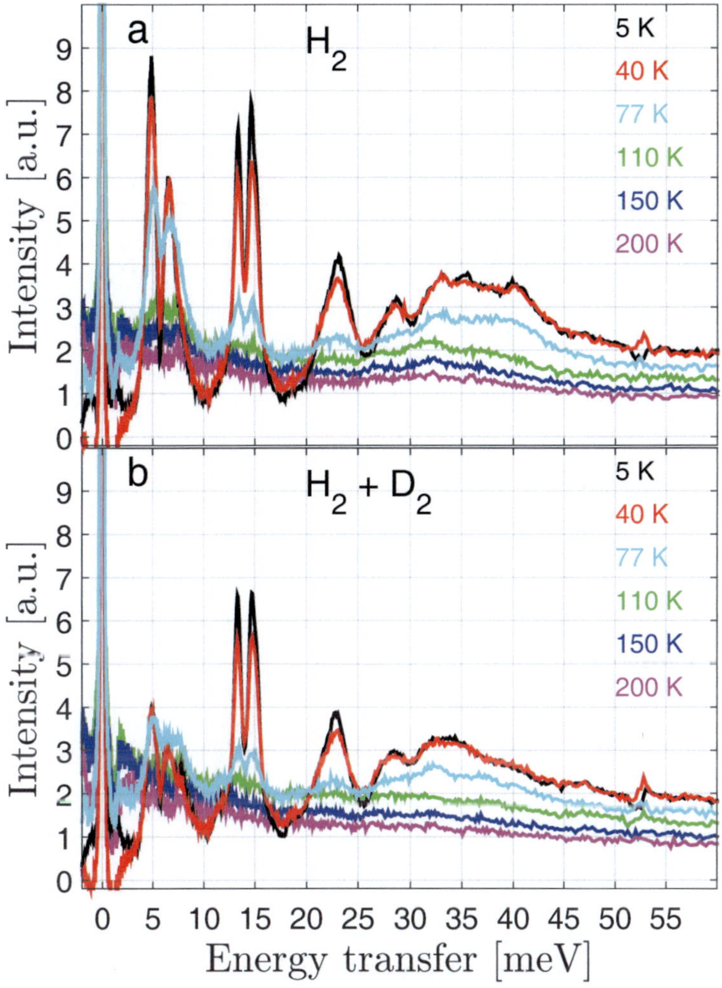

Figure 6.17: a) INS spectra at different temperatures after high-hydrogen loading. b) INS spectra at different temperatures after loading with a mixture of H_2 and D_2.

INS spectra of Cu(I)-MFU-4*l* with high-hydrogen loading shown in figure 6.17a have been recorded at different temperatures. Increasing the temperature

leads to decreasing intensities of the H_2 peaks. At 77 K the second double peak lost more intensity than the first one.
Half of the loaded amount of H_2 has been pumped out at 100 K. Approximately the same amount of D_2 has been added at 70 K. After loading, the sample has been cooled to 5 K. Again, spectra at different temperatures have been recorded which are presented in figure 6.17b. Compared to the measurement with pure H_2, the spectra of the mixture show a lower intensity at the first double peak already at 5 K. These peaks vanish at 110 K completely, whereas in case of only H_2 they are still apparent. After the measurement at 200 K the sample has been cooled again down to 5 K and a spectrum has been recorded. The two spectra obtained at 5 K are shown in figure 6.18a for direct comparison. The only difference in treatment is the heating to 200 K in-between the measurements, but gas was neither released nor added to the sample chamber. Figure 6.18b shows the difference of the two spectra shown in figure 6.18a.

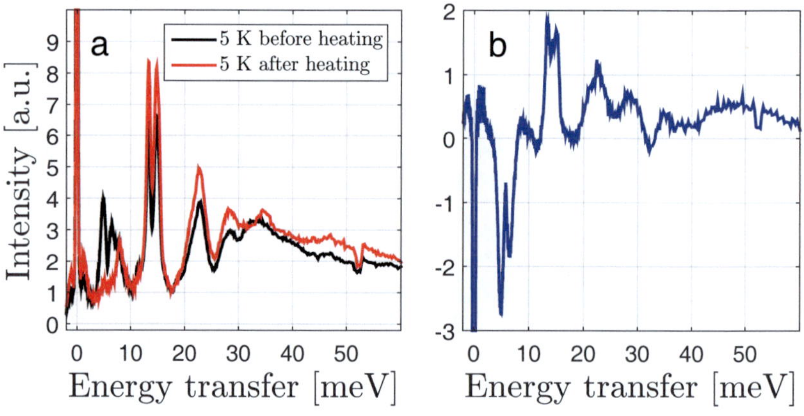

Figure 6.18: a) INS spectra of H_2 in presence of D_2. The black spectrum shows the signal after the direct cooling from the D_2 loading temperature (70 K) to 5 K. The red spectrum shows the measurement at 5 K after a temperature ramp to 200 K has been performed. b) Difference between the red and the black curve in a).

Discussion

Due to the high inelastic neutron cross section of H_2, INS is a convenient technique to investigate molecular excitations of H_2. The corresponding transitions can change significantly for a molecule bound to adsorption sites with an anisotropic potential. The spectra of Cu(I)-MFU-4*l* loaded with two different amounts of H_2 gas are presented in figure 6.16. The molecules are preferentially adsorbed by the strongest adsorption site. Therefore, under low-loading conditions (labeled with 1x loading) only the strong adsorption sites are occupied. If a higher amount is loaded (labeled with 3x loading) additional peaks appear, which show adsorption at the weaker sites.

Adsorption on the strong open metal site of Cu(I)-MFU-4*l*

The double peak at 4.9 meV and 6.5 meV appearing under low-loading conditions represents the H_2 adsorbed to the strong binding site at the open metal center Cu(I). Primarily the spectrum is dominated by the rotational transitions, which are perturbed. Due to adsorption, the rotational constant of H_2 changes, which is $B_{rot} = 7.35\,\text{meV}$ for free H_2. The double peak at 4.9 meV, 6.5 meV and the maximum at 37 meV of H_2 adsorbed on the strong open Cu(I)-site can be explained by using the model of *the perturbed planar rotor in a deep attractive field, the 2-DP type* with a perturbation potential of the form [34, 112]:

$$V(\theta,\phi) = \left[a + \frac{b}{2}\cos 2\phi\right]\sin^2\theta \tag{6.1}$$

θ is the polar angle between the H-H bond and the surface normal, ϕ is the azimuthal angle between the horizontal axis and the projection of the radius vector on the horizontal plane. The rotational barrier parameters a and b give the relative weight of the hindrance in both angles. The energies of the rotational states of H_2 relative to the ground state have been calculated, by using the potential given in eq. 6.1 for the perturbed planar rotor in a deep attractive field. The energy levels for the rotational states of H_2 are shown in figure 6.19 in dependence on the parameter a/B.

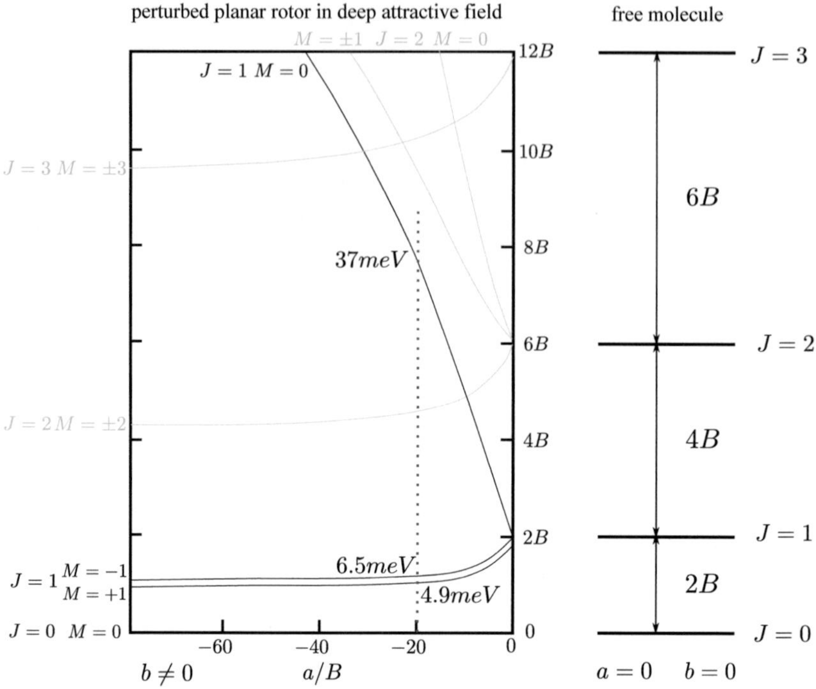

Figure 6.19: Left: Energy levels of the rotational states of H_2 as a planar rotor perturbed in a deep attractive field reproduced from [34]. The dotted line crosses the energy levels, which correspond to the experimentally observed transitions for the strong adsorption site (figure 6.16). Right: Unperturbed rotational energy levels for the free H_2 molecule.

The full black lines show the splitting of the $J = 1$ level and for $b \neq 0$ the degeneracy between $M = \pm 1$ is removed. [60] For $B = 5\,\text{meV}$ and the rotational parameter $a = -100$ (position in the diagram: $a/\text{B} = -20$, figure 6.19 left) the energy levels coincide approximately with the transitions observed by INS at 4.9 meV, 6.5 meV $(J = 0, M = 0)$ to $(J = 1, M = \pm 1)$ and 37 meV $(J = 0, M = 0)$ to $(J = 1, M = 0)$. The dotted line also crosses $(J = 0, M = 0)$ to $(J = 2, M = \pm 2)$, but due to the low coherent cross section between p-H_2 transitions, no line is observed for this energy value.
Combining the model of the perturbed planar rotor (see equation 6.1) with the data obtained from the spectrum (double peak at 4.9 meV and 6.5 meV) allows

to calculate the bond length of H_2 adsorbed on the strong Cu(I)-open metal site. Due to adsorption the molecule elongates to 0.85 Å [111], compared to free H_2 which has a bond length of 0.75 Å [34].

Adsorption on the weak adsorption sites of Cu(I)-MFU-4*l*

The signal dedicated to the strong adsorption site increases slightly in the spectrum taken under high-loading conditions and a second double peak rises at 13.4 meV and 14.7 meV. This indicates that the strong adsorption sites get saturated and weaker sites become populated too. Under the assumption that the effective rotational constant becomes $B = 7.0$ meV, the second double peak can be assigned to scattering, which excites the para-H_2 molecule from the rotational ground state $J = 0$ to the ortho-H_2 state with $J = 1$ by an inelastic collision with a neutron involving its spin flip. This transition has the highest scattering cross section in neutron energy loss spectroscopy. [113] The double peak configuration allows to conclude that the threefold degeneracy of $J = 1$ is lifted to $(J = 1, M = 0)$ and $(J = 1, M = \pm 1)$ by the shape of the potential at the adsorption site (see figure 2.9). [60, 112] Therefore, an intensity ratio of 2:1 for the peaks at 13.4 meV and 14.7 meV would allow to assign them to the $(J = 1, M = \pm 1)$ and $(J = 1, M = 0)$ transition, respectively (see figure 6.20). Given that the $J = 0$ to $J = 1$ transition for free H_2 occurs at 14.7 meV this peak might depict an overlap between the transition of the free and slightly perturbed H_2. [34]
The peak at 23.0 meV is most likely assigned to a combination of a phonon mode and the rotational transition at 14.7 meV. Transitions from the $J = 0$ state to any other excited para-H_2 state have very low cross sections for neutrons, due to the low coherent cross section of H_2 (1.76 barn). [34, 113] Transitions between the ground state and an ortho-H_2 state are controlled by the large incoherent cross section (80.3 barn). A large cross section occurs as well for the combination of a rotational transition and a phonon. [114] Due to the existence of a small amount of ortho-H_2, the signal rising at 28.7 meV belongs to the transition $J = 1$ to $J = 2$ (see figure 6.20).
Both models describe the observed peaks, however, they are based on rough estimations for the rotational constants ($B = 5$ meV and $B = 7$ meV). The real scenario is much more complicated and requires theoretical calculations to fully reveal the physical nature.

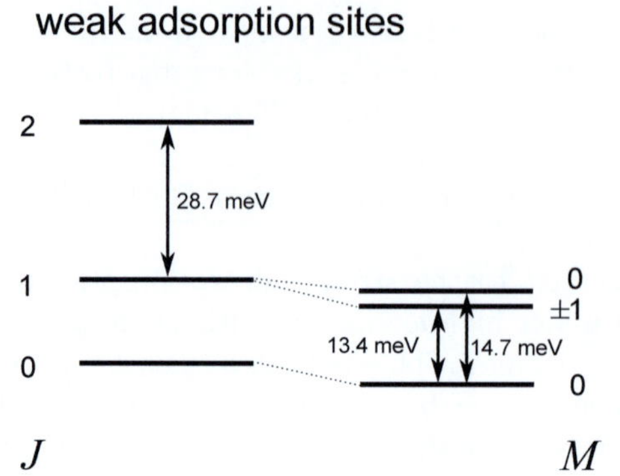

Figure 6.20: Transitions observed in the spectra due to adsorption of the H_2 on the weak adsorption sites.

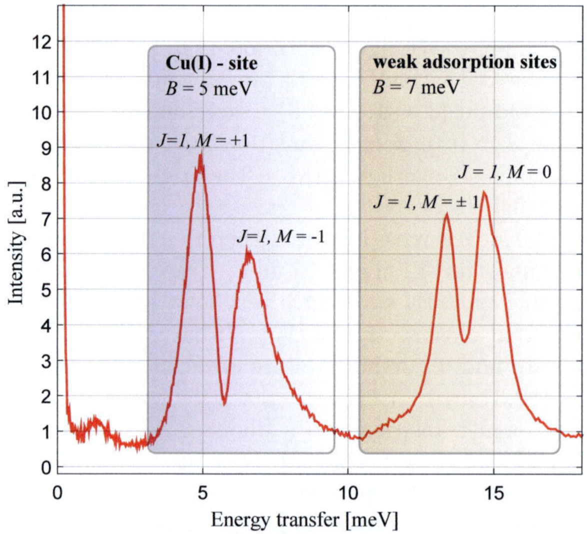

Figure 6.21: Assignment of the double peaks observed in the INS spectrum to the rotational transitions of the H_2 adsorbed at the strong Cu(I)-open metal site and at the weaker sites of Cu(I)-MFU-4*l*. The transitions occur from the ground state ($J = 0$, $M = 0$).

Temperature dependence of the INS spectra

At higher temperatures the INS spectra of H_2 loaded into Cu(I)-MFU-4*l* show a decrease in scattering intensity (figure 6.17a). The different peaks decrease and disappear at different temperatures due to the smaller fraction of adsorbed H_2 and due to a reduced percentage of para-H_2 in the gas. [115] The intensity of the double peak at 4.9 meV and 6.5 meV, which is assigned to H_2 adsorbed at the Cu(I)- open metal site is relatively strong at 77 K and still visible at 110 K. In contrast the double peak at 13.4 meV and 14.7 meV, ascribed to the weaker adsorption sites, decrease faster with temperature and is vanished at 110 K. The same behavior has been observed for the maximum at 23.0 meV, which was explained by a combination of a rotational transition at the weak adsorption site and a phonon mode (see previous discussion).

INS spectra of a hydrogen isotope mixture

For investigating if the observed isotope exchange with TDS can also be detected with INS, half of the H_2 gas has been pumped out at 100 K before adding D_2 to the sample. Approximately the same amount of D_2 has been loaded at 70 K prior to cooling immediately down to 5 K. The first measurement at 5 K already shows a strongly decreased signal at the double peak ascribed to the strong adsorption site (figure 6.17b). The double peak at the weaker adsorption site has almost the same intensity as under loading of pure H_2 and dominates the spectra for 5 K and 40 K. This confirms that H_2 is already partially exchanged by D_2 at the strong adsorption site during the loading at 70 K. Due to its low incoherent cross section of 2.05 barn compared to the 80.27 barn of H_2, adsorption sites occupied by D_2 are not visible in the INS spectra. [59] The spectra taken at different temperatures show that at 77 K the double peak assigned to the weak adsorption sites lost intensity compared to the measurement at 40 K, whereas at the double peak assigned to the strong adsorption site the intensity is comparable between the measurement at 40 K and 77 K. At 110 K the double peak of the strong adsorption sites is lower when both isotopes are loaded (figure 6.17b) compared to its intensity at 110 K, when only H_2 is loaded (figure 6.17a). This shows the exchange of adsorbed H_2 by D_2 at 110 K. After rising the temperature to 200 K and cooling back to 5 K the maxima assigned to the strong adsorption site vanished completely (figure 6.18). Both, the double peak at 13.4 meV and 14.7 meV and the peak at 23.0 meV, which represent interaction of H_2 with the weak adsorption site gain intensity after heating to

200 K. Thus, the H_2 adsorbed at the strong adsorption site have moved to the weak adsorption sites. This has been depicted by subtracting both spectra (see figure 6.18b).

Summary

Owing to its interesting hydrogen isotopes adsorption characteristics, Cu(I)-MFU-4*l* was additionally investigated by INS. Rotational transitions from $J = 0$ to $J = 1$ of hydrogen adsorbed to the strong open metal site appear in the INS spectrum as a double peak at 4.9 meV and 6.5 meV. By introducing a higher amount of H_2 additional peaks appear, which are attributed to rotational transitions of H_2 adsorbed at weaker adsorption sites. Combining the energy values obtained for H_2 adsorbed at the very strong open metal site with the model of the perturbed planar rotor in a deep attractive field, an elongation of the H_2 molecules bond length by 0.1 Å to 0.85 Å could be determined.
Introducing D_2 in addition to the already adsorbed H_2 lead to a decreasing double peak of the very strong adsorption sites, whereas the peaks belonging to H_2 adsorbed to the weak adsorption sites are still relatively high. Due to its lower cross section, D_2 is not visible in the spectrum and, therefore, the reduced signal of the strong adsorption sites shows that already adsorbed H_2 is exchanged by D_2. This is in agreement with the isotope exchange observed with TDS.

6.1.3 Raman spectroscopy

A Raman spectroscopy setup of the group of Prof. Keimer from the Max-Planck-Institute for Solid State research in Stuttgart has been employed for these measurements. This setup has a sample cell especially designed for the investigation of gas adsorption in porous materials. It is designed for pressures up to 25 bar, which for the present experiments have been provided with a gas bottle directly connected to the cell. The pressure is regulated with a pressure transducer. Only half of the cell has been filled with already activated Cu(I)-MFU-4*l*, corresponding to approximately 20 mg. The empty half of the cell was filled with free hydrogen or deuterium gas, whose vibrational transitions thus also could be measured and be taken for reference. Experimental details can be found in chapter 3.3.

Results

Figure 6.22 shows the Q(J)-branch of H_2 gas located around 4155 cm^{-1} and D_2 gas located around 2990 cm^{-1} at different temperatures and pressures. For both isotopes at low temperature only the initial rotational states $J = 0$ and $J = 1$ are populated. Higher rotational states get occupied with rising temperature and at 100 K the Q(2) and the Q(2) and Q(3) transition get visible in the Raman spectrum for H_2 and D_2, respectively. At room temperature the Raman spectrum reveals four Q-lines for H_2 and five Q-lines for D_2, indicating the population up to the levels $J = 3$ and $J = 4$, respectively. The corresponding frequency values are summarized in table 6.3. [116] The Q(0)- and Q(1)-lines are well separated in case of H_2, whereas they overlap for D_2.

vibrational line	Q(0)	Q(1)	Q(2)	Q(3)	Q(4)
$\mathbf{H_2}$ [cm^{-1}]	4164	4157	4146	4128	–
$\mathbf{D_2}$ [cm^{-1}]	2995	2993	2989	2982	2974

Table 6.3: Q-lines of free H_2 and D_2 at 300 K.

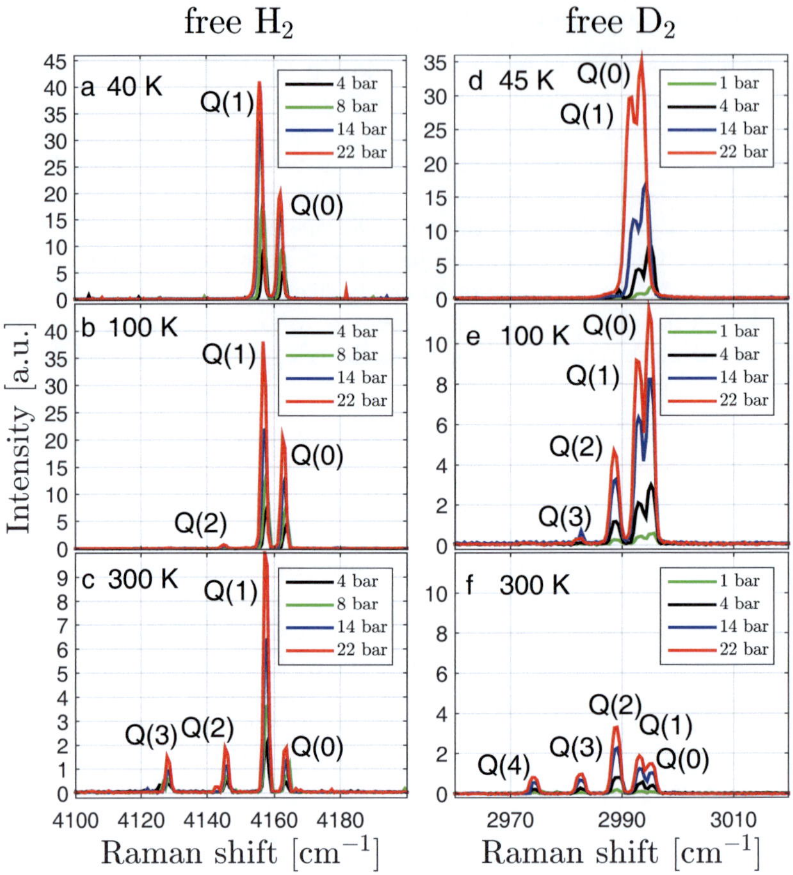

Figure 6.22: Temperature and pressure behavior of the Q- branch of H_2 (left) and D_2 (right).

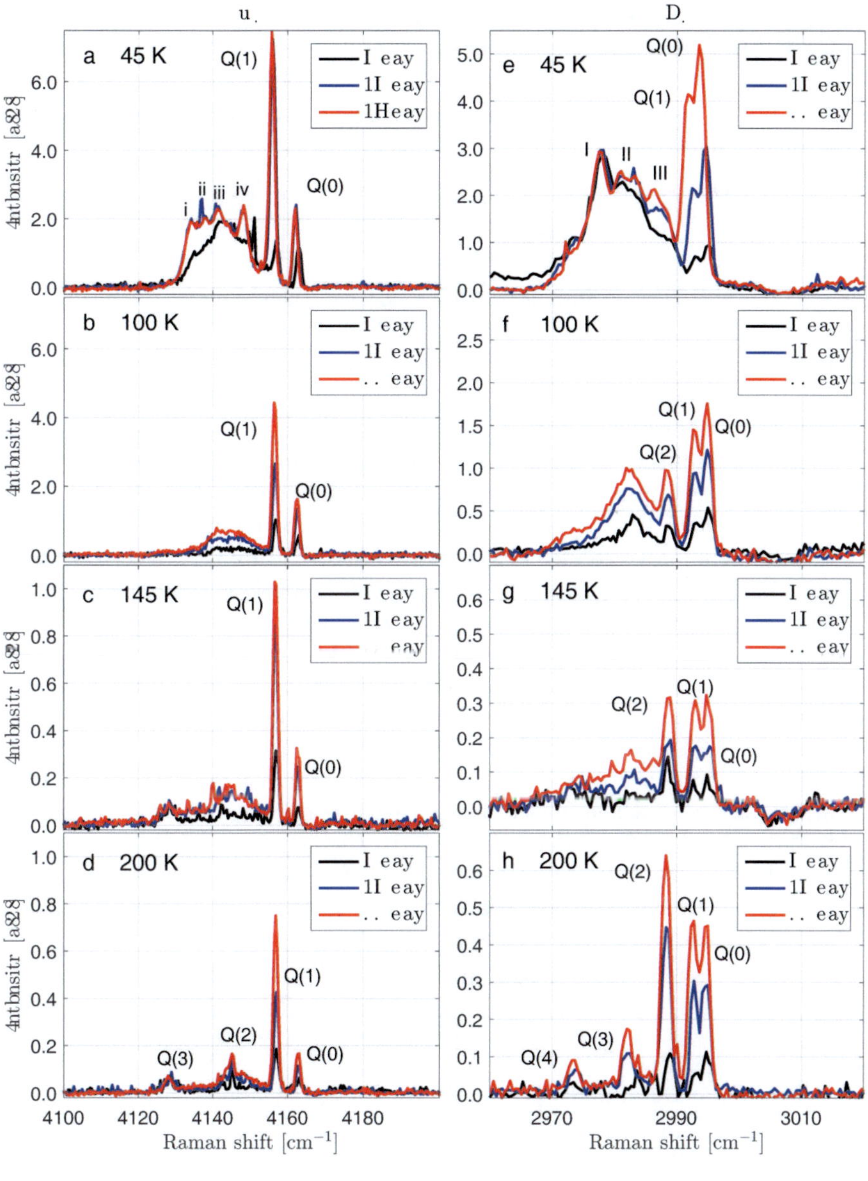

Figure 6.23: Hydrogen (left) and deuterium (right) in Cu(I)-MFU-4*l* at different temperatures and pressures.

Figure 6.23 shows on the right side the Raman spectra of D_2 loaded into Cu(I)-MFU-4*l* at different temperatures and pressures. In the spectrum recorded at 45 K (figure 6.23e), additional to the Q-lines of D_2 gas, maxima are observed with a shift to lower wavenumbers (see table 6.4), which can be ascribed to the adsorbed molecules.

D_2	I	II	III
ν [cm$^-$1]	2978	2982	2987

Table 6.4: Downshifted peaks with respect to the Q-lines of the free D_2.

Peak I remains constant for pressures above 4 bar. Peak II increases if the applied pressure is increased from 4 bar to 14 bar and stays constant for higher pressures, whereas peak III increases continuously with pressure. At 45 K and 4 bar, the intensity of adsorbed D_2 is larger compared to the one for free D_2, which means that most molecules are adsorbed under these conditions (figure ??a). At 100 K, the Raman spectra show still maxima in the range of 2970 cm^{-1} and 2987 cm^{-1}, which means that at this temperature D_2 is still adsorbed. The signal of adsorbed D_2 at 100 K increases continuously with higher gas pressures. Additionally, the Q(2)-line of free D_2 appears in accordance to the spectrum of free D_2 at 100 K (figure 6.22). At 145 K this Q(2)-line increases further relative to the Q(0)- and Q(1)-lines. At 200 K only the Q-lines of the free D_2 are detected indicating that no adsorption occurs.

Figure 6.23 shows on the left side the Raman spectra of H_2 in the material at different temperatures and pressures. A wide signal appears between 4128 cm^{-1} and 4152 cm^{-1}, which is downshifted with respect to the free H_2 lines (figure 6.23a). For 4 bar and 45 K, similarly to the spectrum of D_2, the intensity of adsorbed H_2 exceeds the intensity of the free lines. In the spectra of higher pressures four shifted peaks appear for the adsorbed H_2 and their positions are summarized in table 6.5.

H_2	i	ii	iii	iv
ν [cm$^-$1]	4135	4138	4142	4148

Table 6.5: Downshifted peaks with respect to the Q-lines of free H_2.

In figure 6.24 the spectrum of the adsorbed and free H_2 at 45 K and 18 bar is approximated by six Gauss peaks. By increasing the temperature to 100 K, a downshifted signal is also visible in the spectrum, indicating that adsorption still occurs (see figure 6.23b). A further increase of the temperature decreases the adsorbed amount and gives rise to additional Q-lines of the H_2 gas, as already seen for D_2.

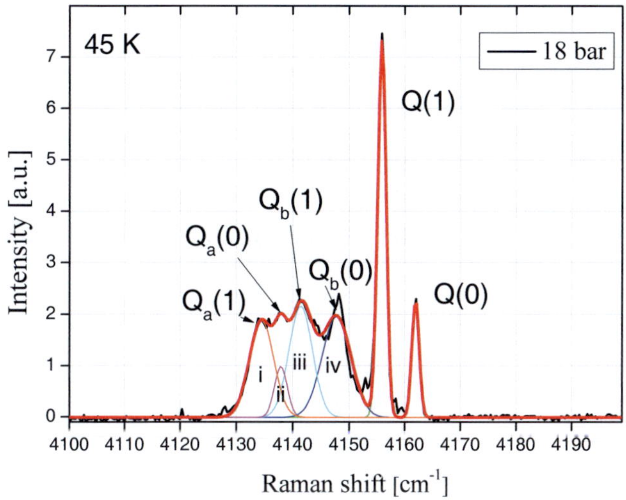

Figure 6.24: Gauss fit of the measured spectrum at 45 K and 18 bar H_2. The red line depicts the sum of the six peaks. The labels a and b emphasize that the Q(0)- and Q(1)-lines of free hydrogen are shifted due to the interaction of the H_2 molecule with different adsorption sites (here named a and b).

Additional to the frequency range discussed above, a wide range between 1000 cm^{-1} and 4400 cm^{-1} has been investigated. However, no other maximum related to H_2 or D_2 has been observed within the resolution of the setup.

Discussion

Raman spectroscopy has been used to study the vibrational frequency of hydrogen isotopes adsorbed in Cu(I)-MFU-4*l* compared to free gas. The shift in stretch frequency provides information about the strength of the interaction energy between the molecule and the binding sites of the material.

Raman spectra of free hydrogen isotopes

First, the Q-lines of free H_2 and D_2 molecules have been recorded at different temperatures and pressures. Under free gas conditions perturbations of the vibrational motions, which depend on gas density and pressure lead to a frequency shift of the Q-lines. Fiutak and Van Kranendonk have shown that the isotropic component of the intermolecular forces is responsible for this shift. [117,118] From the measurements presented in figures 6.22a,b,d,e the Raman shift of the most intense line of free H_2 (Q(1)) and of the most intense line for free D_2 (Q(0)) has been depicted in figure 6.25 in dependence on pressure. The downshift of the Q-lines is linearly increasing for higher pressures.

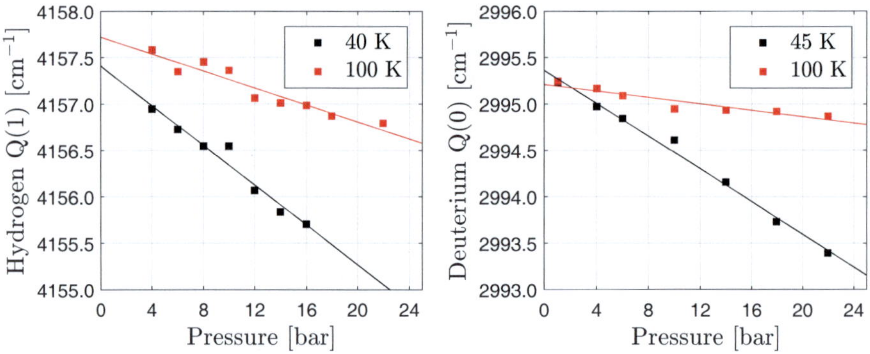

Figure 6.25: Left: Raman shift of the Q(1)-line of H_2 as a function of pressure for 40 K and 100 K. Right: Raman shift of the Q(0)-line of D_2 as a function of pressure for 45 K and 100 K.

The pressure and gas density dependence of the Q-lines can be described by a power series [119–121]:

$$[Q(J)]_\rho = Q(J) + a_J\rho + b_J\rho^2. \tag{6.2}$$

$[Q(J)]_\rho$ is the frequency of the Q-line at the gas density ρ. $Q(J)$ represents the vibrational Raman transition of the isolated molecule (zero pressure limit) in the rotational state J. a_J ($a_J < 0$) and b_J are temperature and J-dependent constants. In agreement with [120,121] a negative value of a_J, i.e. a downshift with increasing pressure is observed. Furthermore, at 40 K the pressure dependence of the shift is more pronounced owing to the higher gas density compared to 100 K. Since a linear pressure dependence is found, the second order term can be neglected in the investigated pressure range.

Raman spectra of adsorbed hydrogen isotopes

For physisorbed hydrogen isotopes, an additional perturbation of the intramolecular potential (V^{Intra}_{H-H}) occurs, which is at least perturbed by three different intermolecular potentials. These arise from the interactions between the adsorbed and gaseous molecules ($V^{Inter}_{H_2,a-H_2,g}$), the adsorbed molecules and the material ($V^{Inter}_{H_2,a-material}$) and between molecules in the adsorbed layer ($V^{Inter}_{H_2,a-H_2,a}$) [122]:

$$\begin{aligned}\mathbf{V}^{total} &= \mathbf{V}^{Intra}_{H-H} + \mathbf{V}^{Inter}_{H_2,a-H_2,g} \\ &+ \mathbf{V}^{Inter}_{H_2,a-material} + \mathbf{V}^{Inter}_{H_2,a-H_2,a}\end{aligned} \tag{6.3}$$

All intermolecular potentials can be described by the Lennard-Jones (6,12) form with r being the distance between two interacting particles (see chapter 2.1):

$$\mathbf{V}^{Inter}_{H_2,a-H_2,a}(\mathbf{r}) = \frac{\mathbf{A}^{H_2,a-H_2,a}}{\mathbf{r}^{12}_{H_2,a-H_2,a}} - \frac{\mathbf{A}^{H_2,a-H_2,a}}{\mathbf{r}^{6}_{H_2,a-H_2,a}} \tag{6.4}$$

The shift of the Q-lines ($\Delta\nu_{H-H}$) depends on the balance of the repulsive and attractive parts of the Lennard-Jones potential of all intermolecular potentials depicted in equation 6.3:

$$\Delta\nu_{H-H} = \frac{\mathbf{K}^{REP}_{H_2,a-H_2,g}}{\mathbf{r}^{12}_{H_2,a-H_2,g}} - \frac{\mathbf{K}^{ATT}_{H_2,a-H_2,g}}{\mathbf{r}^{6}_{H_2,a-H_2,g}} + \frac{\mathbf{K}^{REP}_{H_2,a-material}}{\mathbf{r}^{12}_{H_2,a-material}} - \frac{\mathbf{K}^{ATT}_{H_2,a-material}}{\mathbf{r}^{6}_{H_2,a-material}} + \frac{\mathbf{K}^{REP}_{H_2,a-H_2,a}}{\mathbf{r}^{12}_{H_2,a-H_2,a}} - \frac{\mathbf{K}^{ATT}_{H_2,a-H_2,a}}{\mathbf{r}^{6}_{H_2,a-H_2,a}} \quad (6.5)$$

The intramolecular bond is weakened by the attractive part and causes a frequency downshift, whereas the repulsive part compresses the intramolecular bond and the vibration of the molecule is upshifted, relative to the isolated entity.

For Cu(I)-MFU-4*l* at 45 K, adsorbed D_2 shows vibrational transitions between 2965 cm^{-1} and 2990 cm^{-1} in addition to the free Q(0)- and Q(1)-lines. The downshift reveals that attractive forces between the host and the molecule are dominant. This signal of the physisorbed D_2 can be subdivided into three peaks (figure 6.23e and table 6.4), which indicate that different adsorption sites exist and the frequency is perturbed according to the different strength. Due to the strong overlap of the Q(0)- and Q(1)-lines it can be assumed, that the shifted peaks consist of both lines, which are not separated. The lines of free D_2 have a significantly lower intensity for pressures up to 4 bar compared to the peaks of adsorbed gas. This indicates that up to 4 bar, at 45 K, most D_2 is bound to the framework.
In contrast to the lines of free D_2, which show a downshift with rising pressure, the peaks (I) and (II) do not change their position. This leads to the conclusion that the density of the adsorbed D_2 remains constant in the applied pressure range. Peak (I) has the largest shift from the Q-lines of free D_2, thus, it is related to the strongest observed perturbation of the physisorbed molecules. The constant peak height for pressures of 4 bar and higher indicates that this site is already fully occupied at 4 bar. For the second adsorption site, represented through (II) the saturation occurs at 14 bar. Peak (III), which is denoted to the weakest adsorption site, due to its small downshift, does not saturate within the investigated pressure range up to 22 bar. Taking into consideration that these measurements have been recorded close to the critical temperature of D_2 (38.2 K), peak (III) may be caused by the formation of liquid D_2 with a Raman shift at 2987.99 cm^{-1}. [78,123]

In case of H_2 adsorbed to Cu(I)-MFU-4*l* a downshifted signal relative to the free H_2 lines appears at temperatures below 200 K. At 45 K and pressures of 14 bar and higher, the Raman spectrum between 4128 cm^{-1} and 4152 cm^{-1} can be grouped in four peaks (see table 6.5 and figure 6.24) The Q(0)- and Q(1)-lines of free H_2 are well separated in contrast to the D_2 lines. Thus, the signal from adsorbed H_2 can be assigned to doublets, corresponding to the shift of the Q(0)- and Q(1)-line. The peaks (i) and (ii) correspond to the shifted Q(1)- and Q(0)-line of H_2 adsorbed on the stronger adsorption site and the peaks (iii) and (iv) represent the interaction with the weaker adsorption site.

The observed shifts arise from physisorbed hydrogen isotopes on the weak adsorption sites of Cu(I)-MFU-4*l*. As seen by TDS, the weak adsorption sites of this material are the same as for MFU-4*l*. From adsorption isotherms of H_2 measured up to 20 bar it is known that MFU-4*l* has a high uptake of up to 4 wt% at 77 K. [95] Therefore, the observed shifted signals from the Q-lines can be attributed to adsorption on the weak sites appearing in the TDS spectrum at 30 K and 50 K (see figure 6.14).

Calculations (see figure 6.26) and measurements with INS (chapter 6.1.2) show, that H_2 adsorbed at the strong copper metal site of Cu(I)-MFU-4*l* has a bond length which is elongated to approximately 0.83 Å (INS: 0.85 Å) (free H_2: 0.74 Å) [34]. The elongated bond length should lead to a shift of approximately 800 cm^{-1} and the Q-lines of the adsorbed H_2 then should be visible at 3320 cm^{-1} in the Raman spectrum. This large shift is not visible in the Raman spectra, which were recorded at a wide range of 1000 cm^{-1} to 4400 cm^{-1} and did not show any signal attributable to hydrogen isotopes besides those close to the Q-lines of free H_2. An explanation for this missing, shifted line can be found when the calculations yielding the elongated bond length of 0.83 Å are analyzed in more detail. While bound to the strong adsorption site, the H_2 is found with reasonably high probability in a distance range of 1.4 Å to 2.2 Å to the copper metal site, resulting in a bond length varying from approximately 0.79 Å to 0.88 Å (see figure 6.26). Thus, the frequency shift is not a fixed value of 3320 cm^{-1} but rather a range of values from 2940 cm^{-1} to 3800 cm^{-1}. From these calculations, the line is not expected to be sharp, but smeared out over a large range of wavenumbers and hence of very low intensity, which the used Raman spectrometer was not able to detect. Also, the total number of open metal sites was only 30% of the maximum possible due to imperfections in the synthesis process, which leads to an additional reduction in the signal intensity.

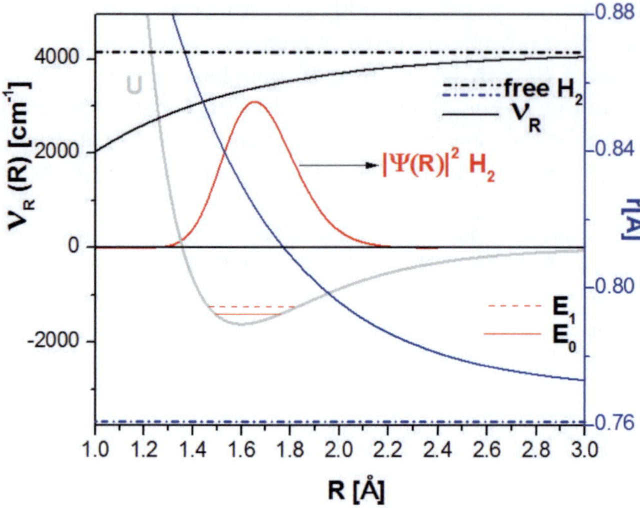

Figure 6.26: The x-axis shows the distance R between H_2 molecule and copper center. The two y-axes show the frequencies of the intramolecular vibration of H_2 (left) and the average intramolecular bond length of adsorbed H_2 (right). Additional curves indicate the Born-Oppenheimer interaction potential U (dimgray) with the zero-point energy E_0 (solid red line) and the first vibrational state E_1 (dashed red line) as well as the ground state probability density $|\psi|^2$ to find the adsorbed hydrogen at a distance R from the copper center (red curve). Dash-dotted lines show the frequency value (black) and intramolecular bond length (blue) of free H_2. [111]

Shift of the Q-lines for hydrogen isotopes

The direct comparison of the hydrogen isotope downshift is possible by plotting the shift relative to the respective Q(0)-line in percent (figure 6.27). For diatomic molecules, the fractional shift of the wavenumber in the vibrational transitions relative to the lines of free gas is mass independent, thus, equal for isotopes. [124, 125] The consistency between the wavenumber shift in percent of H_2 and D_2 shows that the isotopes have been adsorbed at the same sites.

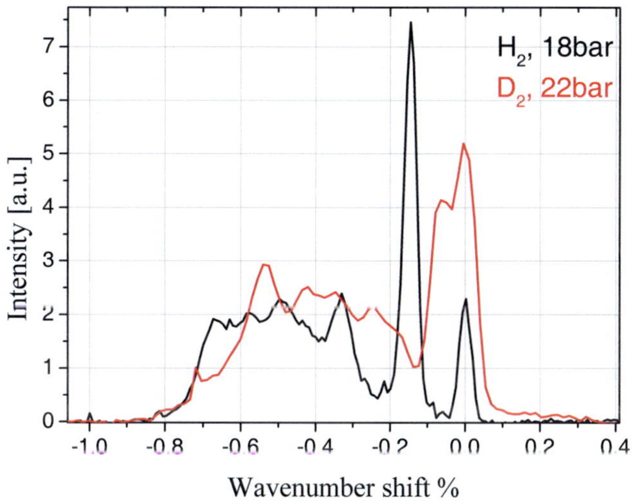

Figure 6.27: The Raman spectra of hydrogen isotopes at 45 K are plotted in percentage wavenumber shift relative to the Q(0)-line.

While the shift of -25% (i.e. about -800 cm^{-1} could not be observed with Raman measurements, a shift of -0.6% (i.e. about -24 cm^{-1}) was found. This shift is too weak to be explained by chemisorption, however, physisorption typically leads to shifts in this range. Therefore, this shift is attributed to interaction of the hydrogen isotopes with the weak adsorption sites of Cu(I)-MFU-4*l* (see figure 6.14). A Raman study by Panella and Hirscher on H_2 adsorbed in single walled carbon nanotubes (SWCNTs) and Cu-BTC presents shifts of -2 cm^{-1}

and 1.5 cm^{-1}, respectively [68]. These values, which the authors attribute to physisorbtion, are more than one order of magnitude smaller than the values determined in this work on Cu(I)-MFU-4*l*. Additionally, the required gas pressures for the observation of this shift are 17 bar in Cu-BTC compared to only 4 bar in Cu(I)-MFU-4*l*, giving further evidence for the stronger adsorption. Another Raman study on the adsorption of H_2 and D_2 in SWNCTs has been conducted by Williams *et al.* with shifts comparable to those seen by Panella and Hirscher (H_2: $<2\,cm^{-1}$; D_2: 0.3-0.5 cm^{-1}) [126]. The authors also explained the observed shifts by physisorbtion. Even higher shifts than those shown in the present work were obtained with IR spectroscopy measurements of MOF-74-Ni (CPO-27-Ni) and MOF-74-Zn showing a shift due to H_2 adsorption at the open metal sites of -120 cm^{-1} and -70 cm^{-1}, respectively. [127,128].

Summary

The Q-lines of free hydrogen isotopes have been measured as a reference for the spectra obtained for hydrogen isotopes adsorbed in Cu(I)-MFU-4*l*. For higher temperatures more lines appear, as higher rotational levels J are occupied. With increasing pressures these lines show a small shift to lower frequencies, which occurs due to small perturbations of the vibration of the molecules due to higher gas densities.

The Raman spectrum of molecular hydrogen isotopes introduced to Cu(I)-MFU-4*l* shows next to the vibrational Q-lines of the free molecules additional red-shifted signals. For both isotopes the same extent in shift percent (-0.6%) from the Q(0)-line occurs, which indicates that the vibration of the molecules is perturbed due to adsorption at the same weak adsorption sites of Cu(I)-MFU-4*l*. Hydrogen adsorbed on the very strong adsorption site of Cu(I)-MFU-4*l* should lead to a much larger shift of about -25%. Calculations have shown that the frequency of the H_2 adsorbed to the open metal site is smeared over a wide wavenumber range and can therefore not be detected within the resolution of the setup.

6.2 Investigation of DUT-9, CPO-27-Ni and Ni-CFA-6 with TDS

6.2.1 Results

DUT-9

The DUT-9 sample has been stored in the argon glovebox and exposed less than 10 min to air during the loading procedure into the TDS device. This MOF has been activated at 393 K for four hours. Figure 6.28 shows the TDS spectra of DUT-9 exposed to 15 mbar pure H_2 or pure D_2 at room temperature. The non-adsorbed gas has been evacuated after cooling to 20 K. All spectra have been recorded with a heating rate of 0.1 K s^{-1} if not explicitly mentioned otherwise.

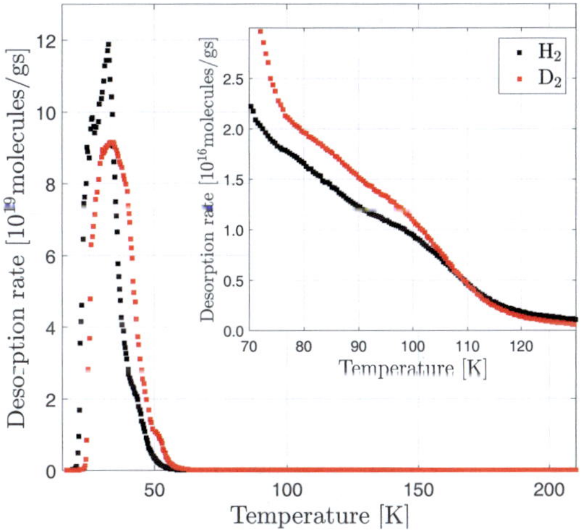

Figure 6.28: TDS spectra of DUT-9 exposed to 15 mbar H_2 or D_2 at room temperature. The inset shows the desorption at higher temperatures magnified. (heating rate: 0.1 K s^{-1})

The D_2 maximum occurs at 33.8 K and is shifted slightly to higher temperatures compared to the sharp maximum of H_2, which is located at 28.9 K. Both gases develop a shoulder around 42 K for H_2 and 51 K for D_2. The inset shows a desorption signal between 73 K and 116 K for both gases.

In figure 6.29 the TDS spectra of DUT-9 after exposure to an equimolar isotope mixture of 10 mbar applied for 10 min at different temperatures is presented. The non-adsorbed gas has always been evacuated at exposure temperature, prior to cooling to 20 K. For better visibility, the maximum at higher temperature is presented in the inset. In contrast to the comparable desorption rates in case of pure gas measurements, the H_2 signal is for all exposure temperatures lower. A maximum between 79 K and 106 K appears only for D_2 when mixtures are applied.

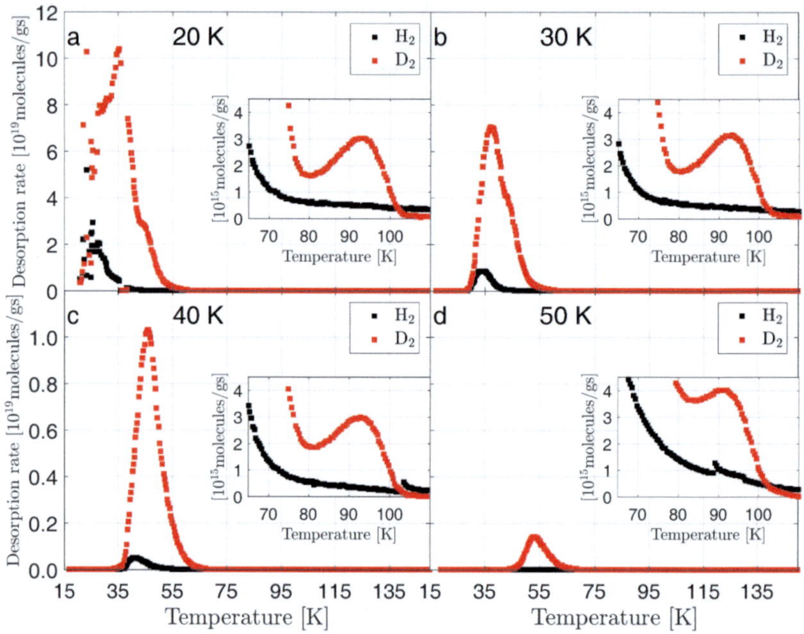

Figure 6.29: TDS spectra of DUT-9 exposed to an equimolar mixture of 10 mbar at different exposure temperatures for 10 min. The inset shows the small maximum occurring at desorption temperatures above 80 K. (heating rate: $0.1\,\mathrm{K\,s^{-1}}$)

Applying 10 mbar at 20 K leads to a high desorption rate around 35 K, which exceeds the maximum range of the mass spectrometer. This explains the interruption in the spectra presented in figure 6.29a. The measurement has been repeated with 5 mbar and a full spectrum without interruption could be recorded, which is presented in figure 6.30a. This measurement shows that the desorption maximum of H_2 appears at about 10 K lower temperatures than the one for D_2. Figure 6.30b shows the maximum at higher desorption temperatures after exposure to 5 mbar and 10 mbar equimolar mixture for 10 min at 20 K. As already observed, a maximum appears only for D_2, which increases slightly for higher pressure. For H_2 the desorption rate is low and at the level of the background signal.

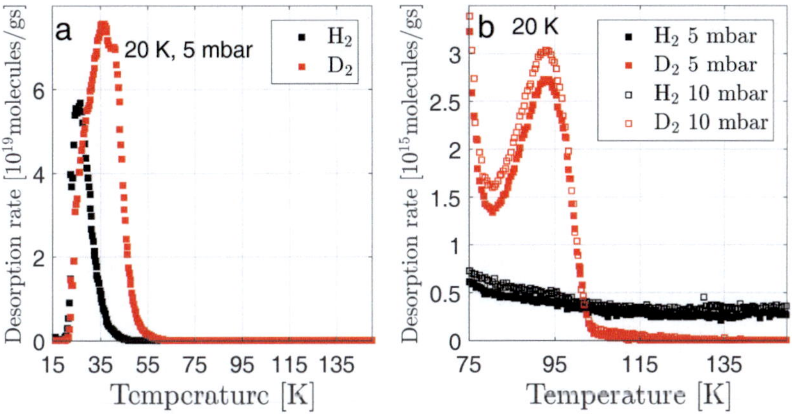

Figure 6.30: a) TDS spectrum of DUT-9 exposed to 5 mbar equimolar mixture at 20 K for 10 min. b) TDS spectrum of the strong adsorption site exposed to 5 mbar and 10 mbar equimolar mixture at 20 K for 10 min. (heating rate: $0.1\,\mathrm{K\,s^{-1}}$)

In figure 6.31 all data from figure 6.29 are presented in one plot with a logarithmic scaling to emphasize that independent of exposure temperature, all spectra coincide on the high temperature flank of the maximum. As a result of evacuation at exposure temperature, the desorption starts at different initial temperatures.

Figure 6.32 shows all data from figure 6.29 presented with a linear scale to depict the descending adsorbed amount for higher exposure temperatures. The inset shows the maximum at higher desorption temperatures, which is constant up to 40 K and increases slightly for both isotopes at 50 K exposure temperature.

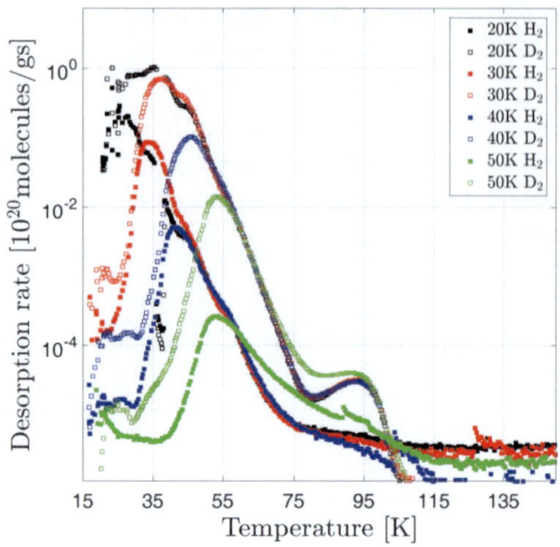

Figure 6.31: Spectra of figure 6.29 presented in one plot with a logarithmic scale. (heating rate: $0.1\,\mathrm{K\,s^{-1}}$)

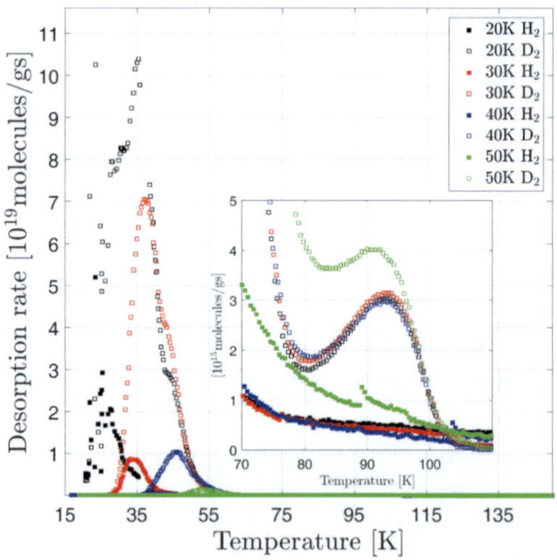

Figure 6.32: Spectra of figure 6.29 presented in one plot. (heating rate: $0.1\,\mathrm{K\,s^{-1}}$)

The selectivity of DUT-9 has been determined from the data presented in figure 6.29 from the desorbed isotopes amount between 15 K and 120 K desorption temperature. Its dependence on exposure temperature is shown in figure 6.33.

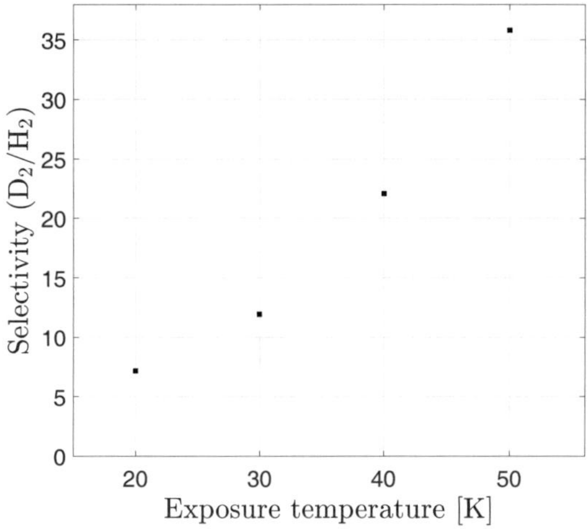

Figure 6.33: Selectivity of DUT-9 in dependence on the exposure temperature.

CPO-27-Ni

CPO-27-Ni has been activated at 423 K for three hours in the TDS device under high vacuum. TDS spectra of CPO-27-Ni exposed to 30 mbar pure H_2 or pure D_2 at room temperature are presented in figure 6.34. The inset shows the TDS spectra with a logarithmic y-axis. The evacuation of the non-adsorbed gas has been done after cooling under gas atmosphere to 20 K and the spectra have been recorded with a heating rate of $0.1\,\mathrm{K\,s^{-1}}$. Two desorption maxima arise for both gases. For each isotope, the maximum at low temperatures is located around 27 K but the location of the maximum at high temperature differs: for H_2 at 75.3 K and for D_2 at 82.4 K. The adsorbed D_2 amount (13.9×10^{21} molecules/g) exceeds the H_2 amount (8.2×10^{21} molecules/g).

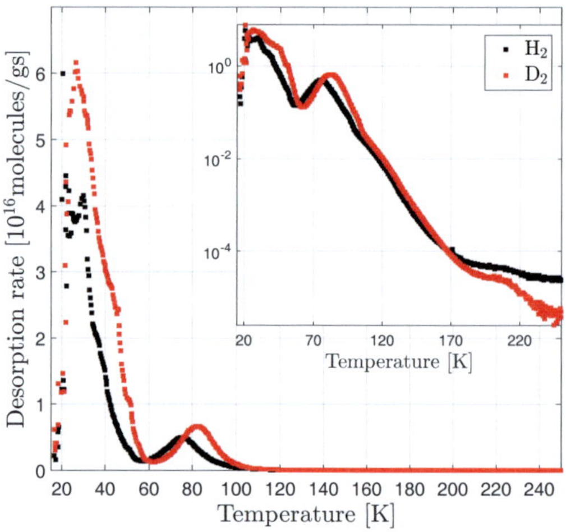

Figure 6.34: Pure gas measurements of CPO-27-Ni. 30 mbar have been applied at room temperature. Inset: the y-axis is scaled logarithmically for better visibility. (heating rate: $0.1\,\mathrm{K\,s^{-1}}$)

Figure 6.35 shows the TDS spectra of CPO-27-Ni exposed to equimolar mixtures of different pressures at different temperatures for 10 min exposure time. The exposure to mixtures leads to much higher maxima for D_2 than for H_2, whereas exposure to pure gas shows comparable maxima for both isotopes. At 60 K exposure temperature the gas adsorbed at the weak adsorption sites (i.e.

desorbing at low temperatures) is removed by evacuation at exposure temperature, thus, only the high temperature maximum remains (figure 6.35g-i).

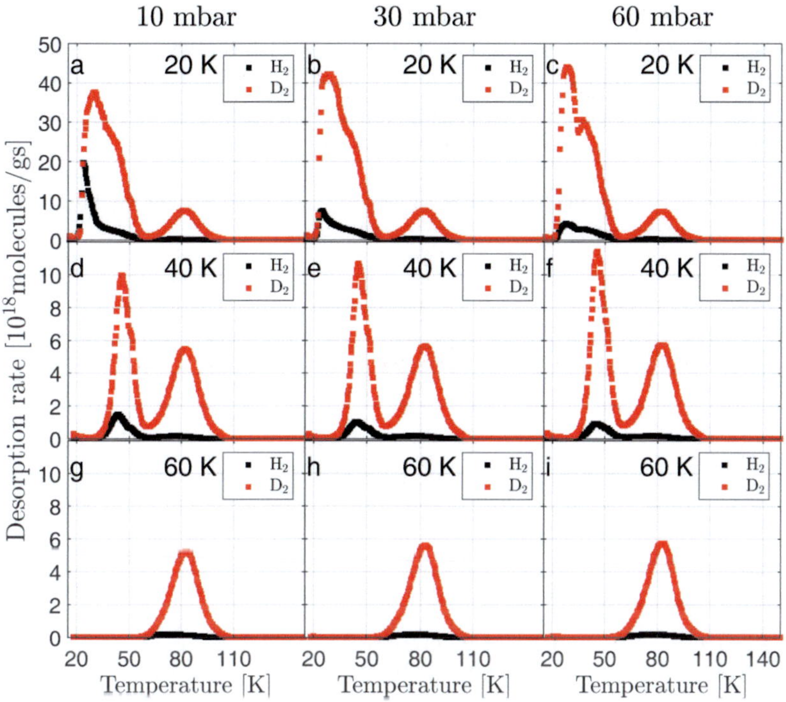

Figure 6.35: TDS spectra of CPO-27-Ni after different temperature and pressures have been applied during exposure. For each temperature, pressure combination the sample has been exposed to an equimolar mixture for 10 min. (heating rate: $0.1\,\mathrm{K\,s^{-1}}$)

Please note, that due to the high adsorbed amount the y-axis for the 20 K measurements differs from that for the other temperatures. The pressure dependence of the spectra, investigated for 10 mbar, 30 mbar and 60 mbar, is very small.

The desorption maximum which rises at 80 K can be assigned to adsorption on the nickel-open metal site. Therefore, the selectivity is presented by using the desorbed gas amount of all adsorption sites (desorption temperature range between 15 K and 130 K) and of only the nickel-site (desorption temperature range between 60 K and 130 K). Figure 6.36 depicts the dependence of the selectivity from the exposure temperatures between 20 K and 80 K for an exposure pressure of 10 mbar. If all adsorption sites are considered the highest selectivity of $S = 18.5$ is reached at 60 K. The selectivity for adsorption on the nickel-adsorption site continuously decreases with increasing exposure temperature from $S = 26.3$ at 30 K to $S = 14.2$ at 80 K. Only the nickel-site contributes to the spectrum at exposure temperatures of 60 K and higher, therefore, the selectivity values at these temperatures are almost identical in both plots.

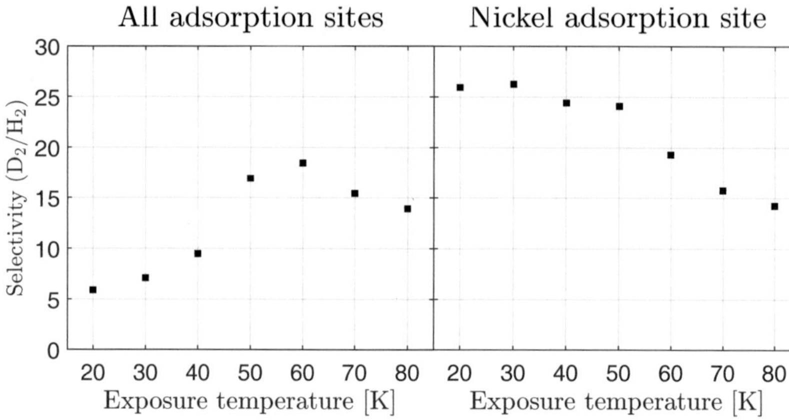

Figure 6.36: Selectivity of CPO-27-Ni in dependence on the exposure temperature, considering all adsorption sites and only the nickel-site. These values are determined from measurements after an equimolar mixture of 10 mbar has been applied for 10 min.

Figure 6.37 depicts the selectivity of CPO-27-Ni at different exposure pressures for different exposure temperatures. Considering the adsorbed amount of all adsorption sites, the selectivity increases for higher exposure pressures at 20 K and 40 K. The selectivity on the nickel-site has no explicit pressure dependence.

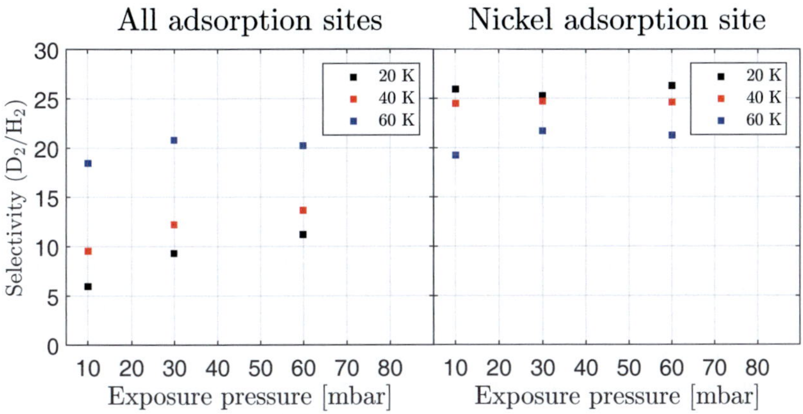

Figure 6.37: Selectivity of CPO-27-Ni in dependence on the exposure pressure, considering all adsorption sites and only the nickel-site. For these values an equimolar mixture has been applied for 10 min at 20 K, 40 K and 60 K.

The extraction of D_2 from an initial low concentration of a H_2:D_2 mixture has been experimentally simulated by a three cycle temperature swing process. In the first cycle, 30 mbar of this mixture with 95% H_2 and 5% D_2 has been applied at 80 K for 10 min. The desorption spectrum shows an enrichment of 86% D_2 and is presented in figure 6.38 as first plot. In the second cycle, 30 mbar of this gas mixture ratio of 14% H_2 and 86% D_2 has been applied again at 80 K for 10 min. The resulting TDS spectrum shows a further enrichment to 97% D_2. In a third cycle, gas with this mixture ratio has been exposed again at 80 K for 10 min, resulting in an identical ratio during the desorption process.

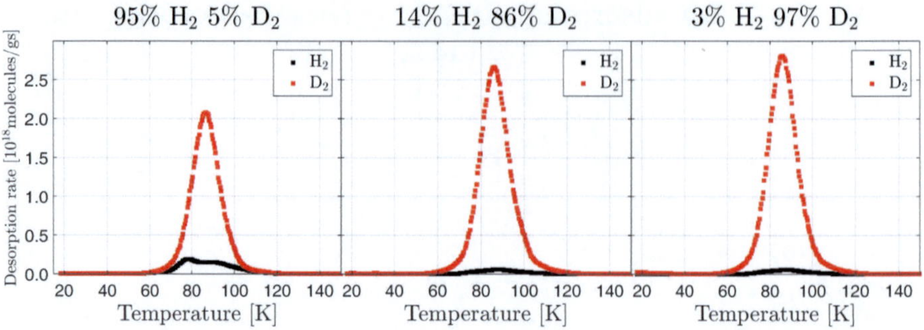

Figure 6.38: For this sequence of measurements, non-equimolar mixtures of 30 mbar have been applied at 80 K for 10 min exposure time. As a starting point, 95% H_2 and 5% D_2 are chosen. For subsequent measurements, the mixture concentrations have been determined from the separation factor of the foregoing measurement. (heating rate: $0.1\,\mathrm{K\,s^{-1}}$)

Ni-CFA-6

The Ni-CFA-6 sample has been activated at 423 K for 2.5 hours before the measurement series has been started. If not mentioned otherwise, the non-adsorbed gas has been evacuated at exposure temperature. Figure 6.39 shows the TDS spectra of Ni-CFA-6 exposed to 10 mbar of H_2 or D_2 at room temperature, recorded with a heating rate of $0.1\,K\,s^{-1}$. The non-adsorbed gas has been evacuated at 20 K.

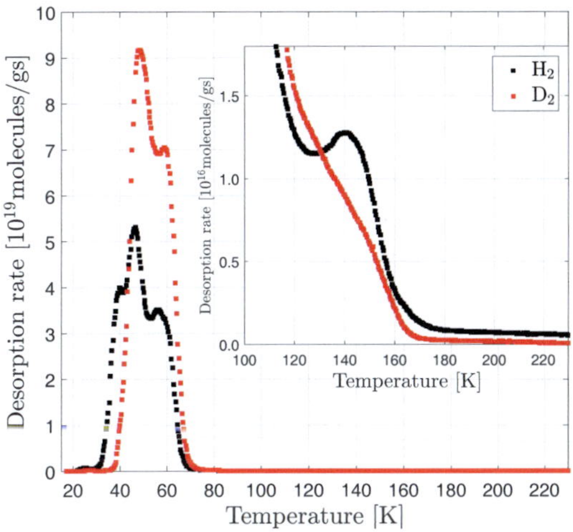

Figure 6.39: Ni-CFA-6 pure gas measurements. 10 mbar of H_2 or D_2 have been added to the sample at room temperature. (heating rate: $0.1\,K\,s^{-1}$)

The desorbed amount of D_2 exceeds the desorbed amount of H_2. Both gases show high desorption signals between 30 K and 72 K. The desorption of H_2 starts already at 30 K, whereas the one for D_2 starts at 37 K. For H_2, the spectrum shows three maxima at 40 K, 47 K and 56 K. D_2 has maxima at 48 K and 59 K. The inset of figure 6.39 shows the desorption rate at higher temperatures. For H_2 a small maximum appears between 125 K and 175 K.

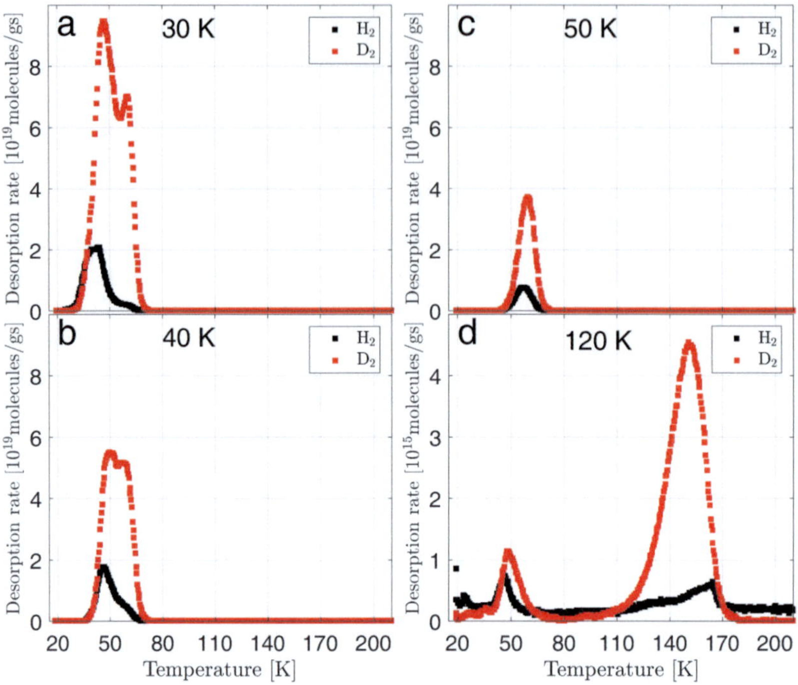

Figure 6.40: TDS spectra of Ni-CFA-6 after applying an equimolar mixture of 10 mbar for 10 min at different exposure temperatures. (heating rate: $0.1\,\mathrm{K\,s^{-1}}$)

Figure 6.40 presents the TDS spectra of Ni-CFA-6 exposed to 10 mbar equimolar mixture at different exposure temperatures for 10 min. Please note that the y-axis has a different scale for the exposure at 120 K. At 30 K and 40 K D_2 still develops two maxima, whereas H_2 develops only a small shoulder around 59 K next to the maximum at 42 K. In figure 6.40c only the higher desorption maxima above 50 K desorption temperature remain due to the evacuation of the non-adsorbed gas at 50 K exposure temperature. In figure 6.40d at 150 K an additional maximum is visible due to the different scale on the y-axis. This maximum appears also for the other three exposure temperatures. However, it is not visible in the chosen scale. The details will be discussed later.

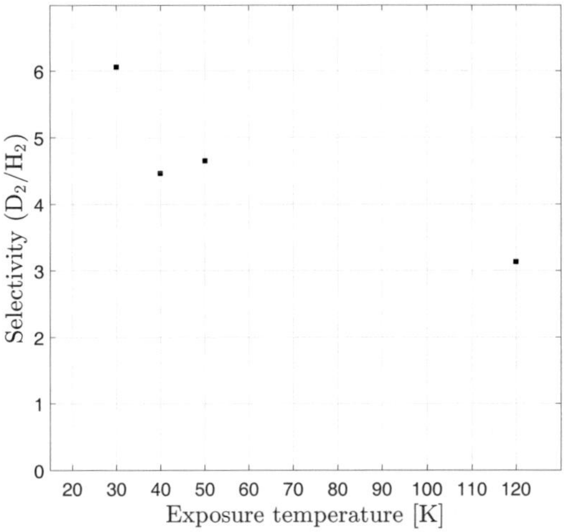

Figure 6.41: Selectivity of Ni-CFA-6 in dependence of the exposure temperature. (heating rate: $0.1\,\mathrm{K\,s^{-1}}$)

The selectivity of Ni-CFA-6 in dependence on the exposure temperature is presented in figure 6.41. Isotopes desorbed in the temperature range between 15 K and 180 K have been taken into account for the calculation of the selectivity. The highest selectivity of $S = 6.1$ is obtained for 30 K exposure temperature. For 40 K and 50 K the selectivity drops around $S = 4.5$. At 120 K exposure temperature, the selectivity reaches $S = 3.1$.

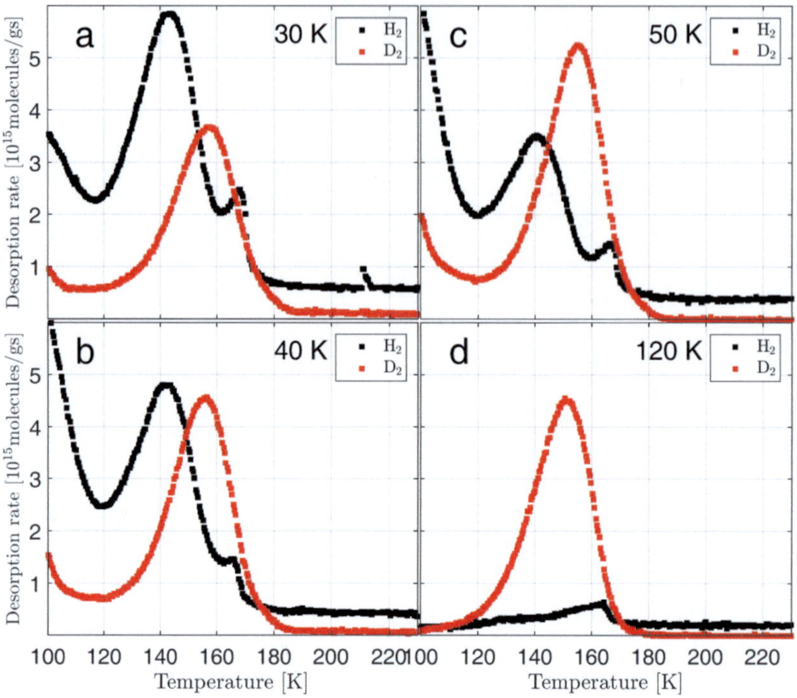

Figure 6.42: TDS spectra of the gas adsorbed on the nickel-site after applying a 10 mbar equimolar mixture for 10 min at different exposure temperatures. (heating rate: $0.1\,\mathrm{K\,s^{-1}}$)

Isotope adsorption at the strong nickel-site has been investigated in detail and the results are presented in the following. Figure 6.42 shows the same data as figure 6.40 with the focus on the high temperature maximum assigned to the nickel-site. The desorption maxima occur for H_2 at 143 K and for D_2 at 157 K. For an exposure temperature of 30 K the H_2 maximum exceeds the D_2 maximum. Increasing the exposure temperature to 40 K leads to a comparable maximum height and for 50 K D_2 exceeds the H_2 maximum. The low H_2 signal at 120 K shows that almost no H_2 is adsorbed. The additional small H_2 maximum, visible around 170 K will occur in all TDS measurements if the scale in this region is magnified. Therefore, this small signal can be considered as background.

The strong nickel-site has also been investigated by applying the 'stepwise loading procedure' at low (30 K) and high (120 K) exposure temperatures. After exposure at 30 K the desorption spectrum consists mainly of the first added isotope, whereas after exposure at 120 K mainly D_2 is desorbed (figure 6.43).

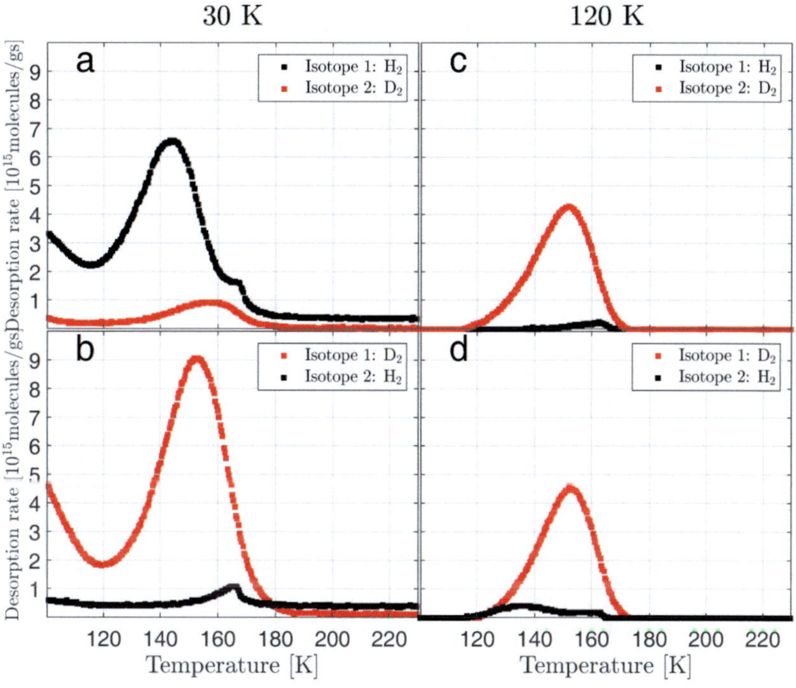

Figure 6.43: Stepwise loading procedure in Ni-CFA-6 at 30 K (left) and at 120 K (right). 5 mbar of each isotope has been loaded for 10 min. (heating rate: $0.1\,K\,s^{-1}$)

6.2.2 Discussion

Three materials containing nickel as open metal sites have been investigated: DUT-9, CPO-27 and CFA-6. The TDS spectra of all materials can be divided in a low temperature region and a high temperature region. The maxima in the low temperature region occur due to desorption from the weak adsorption sites, i.e. linker of the materials, and in the high temperature region molecules desorbing from the strong open metal sites are detected.
For efficient isotope separation through Chemical Affinity Sieving (CAS), strong adsorption potentials are required, which lead to a large difference of the zero-point energies of H_2 and D_2 (see chapter 2.2.3).

Adsorption sites

CPO-27 containing nickel as open metal site is a very well studied sample in the literature. [127, 129–131] The adsorption sites have been determined with neutron powder diffraction by loading different amounts of D_2. The identified sites are presented in figure 6.44. [130] If a low gas amount is loaded, only site I (blue) will be populated, which is the strongest adsorption site in CPO-27. By further loading of D_2, two more adsorption sites are occupied, named site II (yellow) and III (pink). Site I is the strong Ni open metal site, site II is at the triangle of the oxygen atoms of the material and site III is the weakest site, located at the benzene ring. The TDS spectrum of CPO-27 at low exposure temperatures shows all adsorption sites. In a TDS experiment at the lowest desorption temperature the weakly adsorbed molecules get released first. Therefore, for CPO-27 site III appears at 29 K and site II is allocated at 38 K (see figure 6.45). The maximum of the strong open metal site (site I) occurs at 82 K.
The TDS maxima of all three adsorption sites are nearly independent of the pressure applied during exposure (see figure 6.35). This is due to the large channel diameters of the CPO-27 series (11 Å see chapter 4.5), which do not limit the diffusion for the hydrogen isotopes and the adsorption sites can be easily reached.

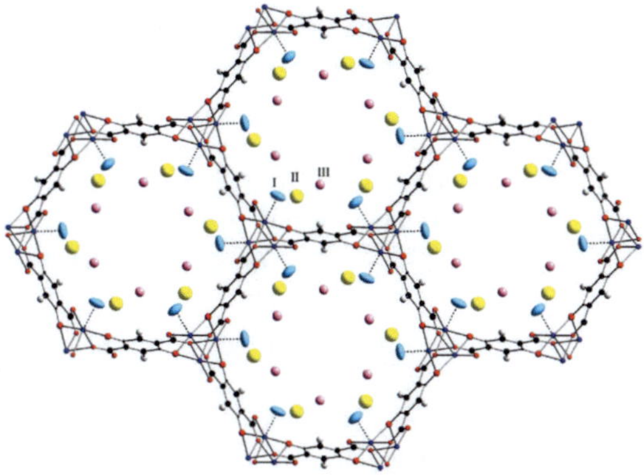

Figure 6.44: The adsorption sites (I blue, II yellow, III pink) of D_2 in CPO-27-Ni determined with neutron powder diffraction. Dark blue: Ni atoms, black: C atoms, red: O atoms, grey: H atoms. [130]

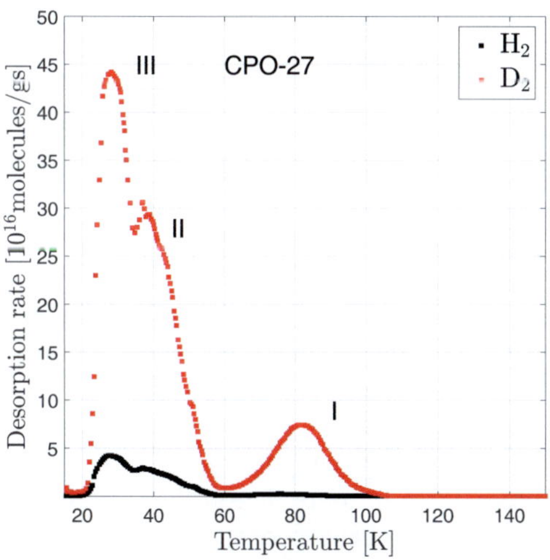

Figure 6.45: TDS spectrum of CPO-27 with nickel as open metal site after exposure to 60 mbar equimolar isotope mixture at 20 K for 10 min. The adsorption sites are denoted in consistence with the denotation in figure 6.44.

Analogues to CPO-27, the TDS spectra of DUT-9 and CFA-6 can be divided in maxima occurring from the weak and maxima occurring from the strong adsorption sites. The TDS spectra of all materials show that the highest amount of gas is adsorbed at the weak sites. The corresponding maxima occur between 21 K-65 K for DUT-9 and 30 K-80 K for CFA-6. In the low temperature region (especially visible for CFA-6, figure 6.39) the desorption signal is built by overlapping maxima, indicating adsorption sites with slightly different potentials.

Hydrogen isotope adsorption on the strong open metal site

Although all materials contain the same metal as strong adsorption site, the TDS spectra differ significantly. Figure 6.46 shows TDS spectra of the three materials measured under identical conditions of 10 mbar equimolar mixtures exposed for 10 min at 30 K (left) and 40 K (right). When mixtures are applied almost only D_2 is adsorbed at the open metal site of DUT-9 and CPO-27, indicating strong Chemical Affinity Sieving (see figure 6.29 and figure 6.35).

	DUT-9	CPO-27	CFA-6
T_{max} H_2 [K]	–	76	143
T_{max} D_2 [K]	93	82	158

Table 6.6: Temperature of the maxima corresponding to the desorbed amount of the strong nickel open metal site of DUT-9, CPO-27 and CFA-6.

Table 6.6 presents the temperatures of the maxima, which correspond to desorption from the open metal sites after an equimolar mixture has been loaded. The temperature difference between the maximum of H_2 and D_2 is smaller for CPO-27 than for CFA-6, but it shows the same tendency: the lighter isotope has a maximum at lower desorption temperature due to its weaker binding to the strong open metal site. CFA-6 at low exposure temperatures adsorbs more H_2 than D_2. This special behavior does not occur for the other two materials and will be discussed later in more detail. The environment of the open metal site differs in all MOFs and this influences the adsorption potential of the metal site. The percentage of adsorbed amount of molecules at the strong adsorption site is given in table 6.7.

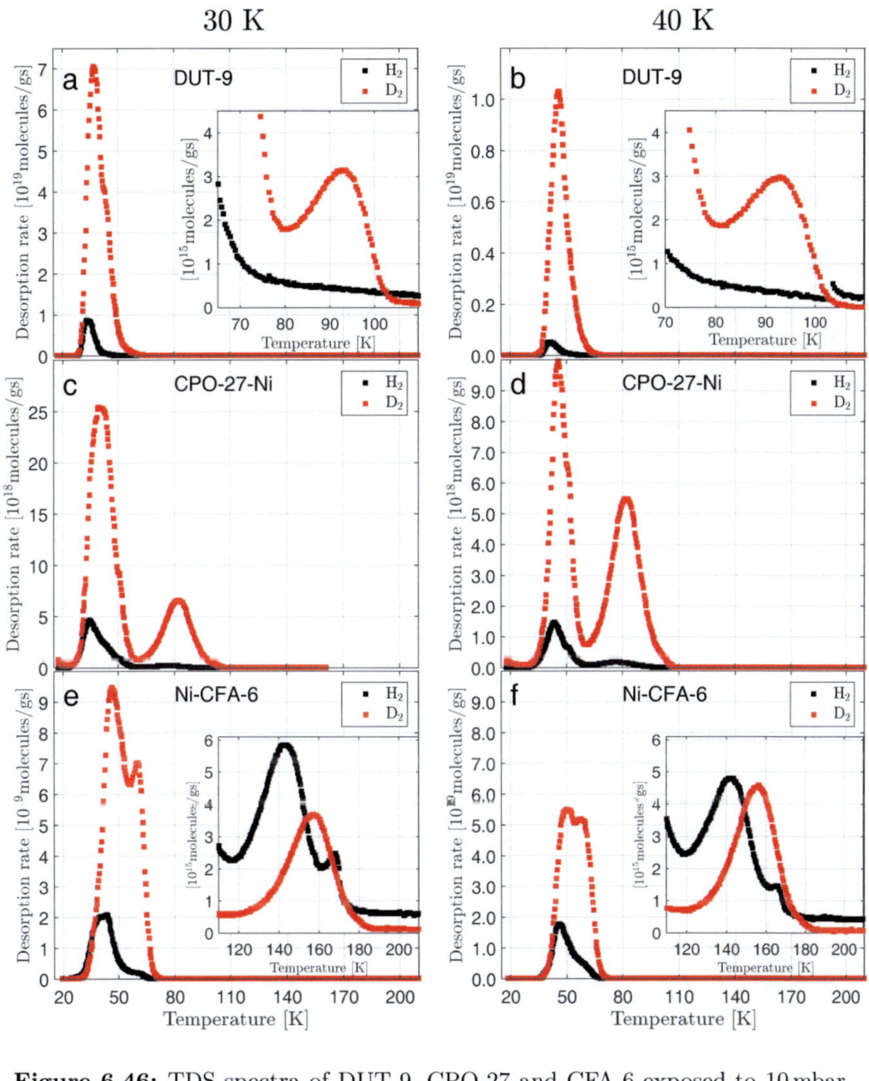

Figure 6.46: TDS spectra of DUT-9, CPO-27 and CFA-6 exposed to 10 mbar equimolar mixutre for 10 min at 30 K (left) and 40 K (right).

adsorbed amount	DUT-9	CPO-27	CFA-6
total [molecules/g]	9.4×10^{21}	6.6×10^{21}	23.8×10^{21}
metal site [molecules/g]	6.1×10^{17}	1.3×10^{21}	3.2×10^{18}
fraction [%]	0.0065	19.3	0.014

Table 6.7: Total adsorbed amount, adsorbed amount at the metal site and the fraction of adsorbed amount at the metal site is given in percent of the total adsorbed amount at 30 K exposure temperature for DUT-9, CPO-27 and CFA-6.

The fraction of gas adsorbed at the open metal site in DUT-9 is extremely low (see table 6.7). DUT-9 has four nickel atoms per cluster (see chapter 4.4) and, therefore, an overall high concentration of metal sites per formula unit. During the synthesis process a part of solvent molecules remain coordinated at the metal sites, which reduces the number of accessible sites. [22] Therefore, the activation procedure has been repeated at 393 K for longer times of 7 h, but no increase in adsorbed amount has been obtained. Even with this extensive activation process only a very low concentration of open metal sites could be reached.

Separation of hydrogen isotopes

The selectivity of DUT-9 is presented in figure 6.33. It increases for higher exposure temperatures from $S = 7.1$ at 20 K to $S = 35.8$ at 50 K exposure temperature. The huge increase in selectivity is accompanied by a big decrease in total gas uptake, i.e. increasing the exposure temperature from 30 K to 50 K reduces the uptake each 10 K step by almost one order of magnitude (see figure 6.31). Owing to the extremely low uptake, the material was not considered for further investigations.
CPO-27 has the lowest maximum temperature, but the high uptake at the strong adsorption site makes this material still interesting for separation of hydrogen isotopes above temperatures of liquid nitrogen (77 K). From literature it is known, that the weak adsorption sites adsorb H_2 with $5\,kJ\,mol^{-1}$ or less [44], whereas for H_2 at the open metal site the adsorption energy is $13.5\,kJ\,mol^{-1}$ [98]. The selectivity of CPO-27 has been determined by considering the adsorbed amount of all adsorption sites and the adsorbed amount of only the open metal sites (see figure 6.36). At 60 K exposure temperature molecules are only adsorbed at the strong open metal site (figure 6.35g-i). With

decreasing adsorbed gas amount at the weak adsorption sites (increasing exposure temperatures) the selectivity increases. As expected for Chemical Affinity Sieving, the higher adsorption energy of the open metal sites leads to a higher difference in zero-point energy for the isotopes and, therefore, the heavier isotope is even more preferential adsorbed. The present results can be directly compared with measurements of the selectivity of CPO-27-Co. The highest selectivity for hydrogen isotopes in CPO-27-Co is also obtained at the exposure temperature, where only the strong adsorption sites are occupied. [44, 132] At 60 K exposure temperature the selectivity is $S = 11.8$ and $S = 18.5$ for cobalt and nickel as open metal site, respectively. The adsorption energy of the strong nickel-open metal site (13.5 kJ mol^{-1} [98]) is higher than the value for cobalt (10.7 kJ mol^{-1} [133]), leading to a higher difference in zero-point energy for CPO-27-Ni and, therefore, a higher selectivity is reached by Chemical Affinity Sieving.

In CFA-6 the highest selectivity of $S = 6.1$ is reached at the lowest exposure temperature, where all weak sites contribute to the separation. At 120 K exposure temperature, the selectivity reaches $S = 3.1$ (figure 6.41). Taking at 120 K only the strong adsorption site into account (maximum between 100 K and 180 K desorption temperature) leads to a slightly higher selectivity of $S = 4.9$. At 120 K exposure temperature, the H_2 desorption is in the same order as the background signal. Therefore, considering the background, the selectivity of the strong open metal site might be much higher.

Isotope exchange at the open metal site of CFA-6

The TDS spectrum of CFA-6 shows that at 30 K exposure temperature, more H_2 than D_2 is adsorbed at the strong adsorption site(figure 6.42a). For 40 K both isotopes are adsorbed approximately to the same extent. At 50 K the adsorbed D_2 amount exceeds the adsorbed H_2 amount and for exposing at 120 K only D_2 is adsorbed. This change in adsorption preference at the open metal site with exposure temperature is further investigated by the method of the stepwise loading. At 30 K exposure temperature (figure 6.43a,b) the first added isotope remains adsorbed and no significant exchange occurs when the second isotope is added. Exposing at 120 K leads always to a high signal for D_2, indicating an exchange of H_2 by D_2 molecules at the strong adsorption sites (figure 6.43c, d). This behavior observed in CFA-6 is very similar to the one observed in Cu(I)-MFU-4*l*. For both materials the desorption maximum assigned to the open metal site occurs at temperatures above 100 K indicating strong interaction. Therefore, at low exposure temperatures the open metal sites are occupied by the isotope reaching the adsorption site first. At this low temperature no further exchange between the adsorbed isotopes at the metal site can occur. The higher H_2 signal at low exposure temperatures after equimolar mixtures have been applied indicates a faster diffusion of H_2 into the pore structure. At intermediate exposure temperatures the occupation reflects the equimolar mixture. Only at higher exposure temperatures, above 100 K, an exchange of the isotopes during the exposure is possible and D_2 with the lower zero-point energy is preferentially adsorbed.

Separation of hydrogen isotopes with the open metal site of CPO-27-Ni

In CPO-27-Ni the desorption maximum assigned to the open metal site is at about 80 K, which is above liquid N_2 temperature (77 K). Furthermore, the open metal site shows a high selectivity of $S = 14$ at 80 K and a high fraction of the total gas adsorbed (19%). Therefore, it gives the possibility for industrial separation with a temperature swing process between 77 K and 140 K. Previously, such a temperature swing process has been experimentally simulated on CPO-27-Co by Oh *et al.* [44]. The same separation cycles have been applied here, starting with a diluted isotope mixture with only 5% D_2. After the first cycle, the desorbed amount of CPO-27-Co and CPO-27-Ni contains 80% and 86% D_2, respectively. Then, after the second cycle, already 97% D_2 are reached for CPO-27-Ni, which cannot be further improved in the third cycle.
For Cu(I)-MFU-4*l*, after the first adsorption/desorption cycle only 81% D_2 have been obtained, compared to 86% D_2 for CPO-27-Ni. However, the adsorbed amount of D_2 in Cu(I)-MFU-4l is 3.3×10^{20} molecules g^{-1} and CPO-27-Ni adsorbs only 5.0×10^{19} molecules g^{-1}. It has to be noted, that the adsorbed amount of CPO-27-Ni does not represent adsorption at all nickel-open metal sites, because evacuation at exposure temperature removes part of the adsorbed D_2 as the maximum appears in the spectrum around 82 K. Furthermore, owing to the strong open metal site of Cu(I)-MFU-4*l* the separation process can be performed at 100 K exposure temperature compared to 80 K for CPO-27-Ni.

Summary

Nickel as open metal site incorporated in three different MOF structures has been studied in this work. The highest selectivities of $S = 18.5$ (at 60 K exposure temperature) and $S = 35.8$ (at 50 K exposure temperature) have been obtained for CPO-27 and DUT-9, respectively. Isotopes desorb from these two materials below 100 K, whereas isotopes desorbing from the open metal site of CFA-6 desorb above 100 K, revealing a stronger adsorption potential. This strong adsorption potential of CFA-6 shows, like for Cu(I)-MFU-4*l*, a preferential adsorption of H_2 from equimolar mixtures exposed at low temperatures. Due to the very strong adsorption sites, which both materials have in common, the isotope exchange has been investigated in CFA-6 as well by introducing the isotopes to the sample one after the other. This procedure confirmed that at low exposure temperatures the first loaded isotope remains adsorbed and no exchange happens, whereas at higher exposure temperatures D_2 loaded secondly exchanges the already adsorbed H_2.

7 Conclusions

In this work, the adsorption and separation of hydrogen and helium isotopes in selected metal-organic frameworks (MOFs) has been investigated. In MOFs, isotope separation can be realized by two principles, either by geometrical confinement in small pores, i.e. Quantum Sieving, or by strong adsorption sites, i.e. Chemical Affinity Sieving. Owing to the lower zero-point energy of the heavier isotope, it penetrates easier and thus faster into structures with pore diameters slightly larger than the kinetic diameter of the isotopes, which leads to Quantum Sieving. Chemical Affinity Sieving is based on the stronger interaction of the heavier isotope with the adsorption site due to its lower zero-point energy. In contrast to Quantum Sieving, the materials selected to investigate the influence of strong adsorption sites have much larger pores than the kinetic diameter of the isotopes. MOFs with small pores or strong adsorption sites have been chosen for studying these sieving principles. The samples have been exposed to an equimolar isotope mixture and the adsorbed quantity of each isotope is detected by low temperature thermal desorption spectroscopy (TDS). The ratio of the desorbed amount of isotopes leads directly to the separation factor, called selectivity.

Adsorption and separation of isotopes in small pore materials

For Quantum Sieving of hydrogen (kinetic diameter 2.89 Å [100]) and helium (2.60 Å [104]) isotopes, several MOFs have been chosen from the IFP and MFU-4 series. IFP-1, -3, -4 and -7 form a hexagonal framework with one dimensional channels. Inside the channels different molecules or atoms restrict the channel diameter periodically to apertures of 1.7 Å (IFP-4), 2.1 Å (IFP-7), 3.1 Å (IFP-3) and 4.2 Å (IFP-1).

Further, two materials have been investigated from the isoreticular MFU-4 series, Cl-MFU-4 and Br-MFU-4. This series has a three dimensional pore structure with two differently sized alternating pore types, separated by an aperture. In the case of chlorine, the aperture diameter is 2.52 Å, whereas for the larger bromine it has a diameter of 2.33 Å.

- The IFP-series can be divided into materials with pore diameters larger (IFP-1 and IFP-3) and smaller (IFP-4 and IFP-7) than the kinetic diameter of the hydrogen isotopes. From a geometrical point of view no gas should penetrate into IFP-4 and IFP-7, but due to the temperature dependent flexibility (also called gate-opening effect [102]) of the apertures penetration becomes possible with increasing temperature.
  The selectivity of the IFP- materials is below $S = 4$ and shows no explicit dependence on exposure time. For IFP-4, with the smallest pore aperture, the selectivity has a constant value of $S = 2$ for different exposure times and pressures, which can be explained by the channel structure of the IFP-samples. Due to the relatively small cavities between the apertures of IFP-4, molecules in the channels cannot pass each other, which leads to a chain-like filling of the channels. Therefore, no time dependence is observed, since the Quantum Sieving effect occurs only at the outermost pore aperture.
  A previous work on hydrogen isotope separation in Cl-MFU-4 found a much higher selectivity of $S = 7.5$ and ascribed it to Quantum Sieving. [102] Cl-MFU-4 is built by an alternation of large and small pores, which are connected by small apertures. In the large pore the isotopes can pass each other and each penetration into the neighboring pore leads to a further sieving step. Therefore, a material in which large cavities and small apertures alternate, is of great advantage for the efficiency of Quantum Sieving.

- Regarding the successful separation of hydrogen isotopes in the bimodal pore structure of the Cl-MFU-4 sample, helium isotope separation has been investigated in Br-MFU-4 and Cl-MFU-4. The apertures of Br-MFU-4 (2.33 Å) and Cl-MFU-4 (2.52 Å) are smaller than the kinetic diameter of the helium isotopes, but, due to the temperature dependent opening of the aperture, penetration into the material is here also possible. For Cl-MFU-4 the difference between the pore aperture at zero-temperature and the kinetic diameter of helium is very small and, therefore, helium penetrates into the structure even at the lowest investigated exposure temperature of 20 K and the so called gate-opening effect [102] is not observed. For Br-MFU-4 though, the highest adsorbed helium amount is observed at 50 K exposure temperature, while no penetration is observed at 30 K exposure temperature and, thus, the gate-opening effect occurs in this material at higher temperatures.

In this work, the highest selectivity obtained for helium isotopes is $S = 2.3$ at 30 K with Cl-MFU-4. For comparison, the highest selectivity reached for hydrogen isotopes in this material is $S = 7.5$ at 60 K. [102]. This factor of three is mainly a result of the three times higher difference in zero-point energy of D_2 and H_2 compared to that of ^{4}He and ^{3}He. The separation of helium isotopes has not been investigated with Chemical Affinity Sieving, as several test measurements with the strongest available open metal site (Cu(I)-MFU-4*l*) yielded no desorption maximum.

- The desorption energy of ^{4}He and hydrogen isotopes has been determined to 5.1 kJ mol^{-1} and 6.6-7.0 kJ mol^{-1}, respectively, for IFP-4 with TDS measurements performed with different heating rates. The results obtained in this work for hydrogen isotopes are in accordance with the value of 6.9 kJ mol^{-1} [93] obtained from conventional adsorption isotherms of H_2. The slightly lower value for ^{4}He compared to hydrogen isotopes results in a TDS maximum at 103 K ^{4}He, which is slightly lower than the maxima of hydrogen isotopes (115 K) due to the smaller kinetic diameter of helium and its faster passing through the pores.

Adsorption and separation of isotopes in materials with open metal sites

For Chemical Affinity Sieving, MOFs with much larger pores (6-30 Å) but strong open copper and nickel centers as part of the structure have been selected. The strongest open metal site has been found in Cu(I)-MFU-4*l*, thus, this material has been deeply investigated by thermal desorption spectroscopy, Raman spectroscopy and inelastic neutron spectroscopy. In addition, three materials with nickel as open metal site: DUT-9, CPO-27 and CFA-6 have been investigated by TDS.

- The TDS spectrum of Cu(I)-MFU-4*l* shows desorption maxima for H_2 and D_2 at 166 K and 181 K, respectively, which are the highest desorption temperatures obtained to our knowledge. These maxima can be assigned to the adsorption on the very strong copper open metal site, where the difference in zero-point energy between the isotopes is large. Therefore, the heavier isotope is stronger bound and is released at higher desorption temperatures. By applying different heating rates in the TDS measurements, the desorption energy of H_2 and D_2 at the copper-open

metal site has been determined to 24.8 kJ mol^{-1} and 33.2 kJ mol^{-1}, respectively, which are in agreement with values known from literature for this material. [19,111]. These values are up to now the highest known for hydrogen isotopes adsorbed in MOFs.
INS spectra were measured at Oak Ridge National Laboratory at the new spectrometer VISION with a high resolution of >2 meV. A double peak at 4.9 meV and 6.5 meV in energy transfer, which can be assigned to the rotational transition $J = 0$ to $J = 1$ of the H_2 strongly bound to the copper open metal site has been detected. This double peak can be explained by the model of the perturbed planar rotor in a deep attractive field. With this model, the bond length of the adsorbed H_2 molecule is determined to 0.85 Å. Thus, the H_2 molecule elongates compared to a free H_2 molecule, by 0.1 Å, i.e. 13%.
Raman measurements on H_2 and D_2 show a clear downshift of the Q-lines at temperatures below 145 K, corresponding to attractive interaction of the molecules with the weak adsorption sites of Cu(I)-MFU-4*l*. The intensity of this signal decreases for higher temperatures, but is still visible at 100 K in agreement with hydrogen adsorption isotherms performed in earlier work. [95] The strong expected shift of the Q-lines, due to adsorption of hydrogen isotopes at the copper-site has not been detected. This is explained by calculations, which show that the signal is smeared out over a wide frequency range. [111]

- Isotope separation is directly studied with TDS after loading equimolar isotope mixtures at different exposure temperatures. At 20 K and 30 K, H_2 with the higher diffusion rate reaches the strong adsorption sites faster. Therefore, the adsorbed amount of H_2 at the open metal sites exceeds the adsorbed amount of D_2. At exposure temperatures between 40 K and 77 K, both isotopes are equally adsorbed. Above 77 K, the amount of adsorbed D_2 exceeds the amount of H_2. This has been observed for the first time. It has been further investigated by the stepwise isotope loading procedure, for which the isotopes are loaded one after the other to the sample at different exposure temperatures. Up to 77 K, the first added isotope remains adsorbed and no maximum is observed for the second loaded isotope. Above 77 K exposure temperature, the adsorbed D_2 amount always exceeds the adsorbed H_2 amount, even if it is loaded secondly. This confirms the exchange of already adsorbed H_2 by D_2 at high exposure temperatures. Therefore, if an equimolar mixture is added

at low exposure temperatures (20 K, 30 K) the thermal energy needed for the exchange of H_2 by the later arriving D_2 is not provided and leads to a high adsorbed amount of the faster diffusing H_2.
The isotope exchange has also been investigated with INS. After loading D_2 to Cu(I)-MFU-4*l*, which was already under H_2 atmosphere, the strong double peak belonging to gas adsorbed at the open metal site decreased. Due to the lower incoherent cross section of D_2 (2.05 barn cf. H_2 80.3 barn [34]), it is not visible in the INS spectrum and, therefore, the observation of decreasing peaks is ascribed to an exchange of the already adsorbed H_2 by D_2. After heating to 200 K the double peak vanished, indicating that only D_2 is adsorbed. The intensity of the double peak assigned to adsorption at the weak sites is still relatively high, which means that H_2 is adsorbed at the weaker sites.
The highest selectivity of Cu(I)-MFU-4*l* has been obtained at an exposure temperature of 100 K. For longer exposure times the selectivity increases from $S = 10.8$ to $S = 25.2$ for 5 min and 90 min, respectively, which clearly shows that the isotope exchange is time dependent.

- Three MOFs containing nickel as open metal site have been investigated. While DUT-9 and CPO-27 show desorption maxima of the strong adsorption site below 100 K, for CFA-6 the respective maxima occur above 100 K. The adsorbed amount of D_2 always exceeds the adsorbed amount of H_2 for the maxima below 100 K. The highest selectivities of $S = 35.8$ and $S = 18.5$ have been obtained for DUT-9 exposed at 50 K and CPO-27 exposed at 60 K, respectively. However, the high selectivity reached for DUT-9 is accompanied with a huge decrease in total gas uptake.
At low exposure temperatures, the open metal site of CFA-6 adsorbs a higher amount of H_2 than of D_2. With rising exposure temperature a change from preferential H_2 adsorption to preferential D_2 adsorption occurs. The strong nickel adsorption site of CFA-6 has also been investigated with the stepwise loading procedure. Indeed, the exchange of H_2 by D_2 at high exposure temperatures, firstly detected at the strong open metal site of Cu(I)-MFU-4*l*, has been observed in CFA-6, too. Therefore, all materials with very strong adsorption sites investigated in this work show an isotope exchange at exposure temperatures above 90 K. For adsorption sites appearing in the spectrum below 100 K, the thermal energy required for an isotope exchange is already provided at low exposure temperature of 20 K, which leads to high selectivity values.

Comparison between Quantum Sieving and Chemical Affinity Sieving

In a direct comparison of investigations on Quantum Sieving (QS) and Chemical Affinity Sieving (CAS) performed in this work, CAS in general gives the better selectivities and, hence, is more suitable for isotope separation. For QS, the required pore apertures of the used MOF must be slightly larger than the kinetic diameter of the isotopes to be separated. When the pores are too large, no separation by QS is possible as both isotopes can enter the pores without hinderance and when the pores are too small none of the isotopes is adsorbed. Occurring flexibilities of the pore apertures and, therefore, alternations in diameter can have additional, unfavorable influences on the separation capabilities of a MOF. In general, MOFs with strong adsorption sites have larger pore diameters compared to the kinetic diameter of the hydrogen isotopes and no geometrical confinement occurs. Strong adsorption sites, like unsaturated metal centers have been already implemented successfully in many MOFs, which seems to be uncomplicated from the point of view of synthesis.
Helium is chemical inactive and hence separation in this work could only be performed with QS, but not with CAS. After applying ^{4}He to the material with the strongest available open metal site (Cu(I)-MFU-4*l*) no desorption maxima appeared in the TDS spectrum, due to the weak van der Waals interaction of helium with the strong adsorption site.
Figure 7.1 shows the highest obtained selectivities (here determined from equimolar hydrogen isotopes mixtures, see eq. 3.1), plotted versus the exposure temperature. This plot visualizes that selectivies obtained by Chemical Affinity Sieving (red) are superior to those obtained by Quantum Sieving (blue).
Quantum Sieving is strongly dependent on the pore aperture diameter and also on the design of the framework. When isotopes cannot pass each other in channel-like structures, the sieving effect happens only at the outermost aperture of the material and once the isotopes have entered, no further sieving occurs inside of the material. For efficient Quantum Sieving usually temperatures below 77 K are required, which therefore, excludes possible cooling with the low cost liquid N_2.

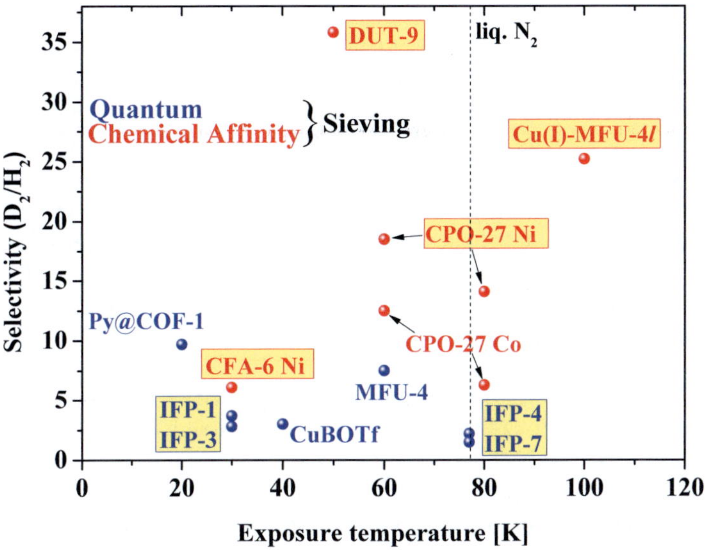

Figure 7.1: Highest selectivities for different porous frameworks obtained after exposure to equimolar hydrogen isotope mixtures at different exposure temperatures. Selectivities obtained with Quantum Sieving and Chemical Affinity Sieving are presented in blue and red, respectively. Materials investigated in this work are highlighted with a yellow background, the values of the other materials are taken from [134].

Metal-organic frameworks with strong open metal sites can be used well above liquid N_2 temperatures for efficient isotope separation. The most suitable material for this task possesses a high selectivity value and a high exposure temperature, which is found in this thesis in Cu(I)-MFU-4*l* (figure 7.1). Therefore, hydrogen isotope separation at the strong open metal site of Cu(I)-MFU-4*l* has been investigated by experimental reproduction of a temperature swing process, which can be performed between 100 K and 220 K. A hydrogen isotope mixture with only 5% D_2 has been introduced for 10 min at 100 K to Cu(I)-MFU-4*l*. Already after the first adsorption/desorption process, a concentration of 81% D_2 has been obtained in the desorbed amount. This isotope

mixture was used for the second adsorption/desorption cycle resulting in a D_2 concentration of 93%. The difference of the zero-point energy at the strong adsorption site allows for the capture of the heavier isotope from a mixture with low concentration. This is ideal for hydrogen isotope separation, as the less abundant isotope, D_2, can be efficiently enriched within only a few cycles and at temperatures well above that of liquid nitrogen and thus at low cost.
The highest selectivity at 100 K obtained in this thesis for Cu(I)-MFU-4*l* is $S = 25.2$. This high selectivity, that was reached even above liquid nitrogen exposure temperature, gives an impressive demonstration of the capabilities of CAS and is to our knowledge state-of-the-art. This work paved the way for CAS, which now raises the question for cost and feasibility of potential applications on industrial scales. The quantities of MOF used in this work never exceeded a few milligrams, however, for industrial applications, an amount on the order of kilograms will be required. The chemicals for one kilogram of Cu(I)-MFU-4*l* cost, when bought in typical quantities used in a chemical laboratory, approximately 60000€. [135] On an industrial scale, the used chemicals will probably be much cheaper.
The obtained results show the potential of MOFs with strong adsorption sites for efficient isotope separation. Further improvements on MOFs is expected, as in Germany alone almost 40 groups are active in this field. The industrial interest in MOFs for different applications like purification of gases, catalysis or gas storage suggests that a cost-efficient production on a large scale, as already performed for several MOFs, will be possible for more materials, which now are limited to a laboratory scale.

8 Zusammenfassung

Einleitung

Die Gewinnung der hoch gefragten Isotope Deuterium und Helium-3, aus den am häufigsten vorkommenden Elemente im Universum, Wasserstoff und Helium stellt immernoch eine große Herausforderung dar. Die natürliche Häufigkeit von Deuterium beträgt 0,015% [1] und Helium-3 ist mit 0,00013% [2] sogar noch seltener. Deuterium wird in vielen industriellen und wissenschaftlichen Anwendungen benötigt, wie zum Beispiel als Moderator in Reaktoren, für Neutronenstreuungsexperimente und als Tracer in der Chemie und Biologie. Helium-3 ist hoch gefragt für den Einsatz in Neutronendetektoren [3] oder in Mischkryostaten [4, 5]. Deuterium und Helium-3 bilden zusammen den Brennstoff für Fusion. Der erst kürzlich gelungene Durchbruch in der Entwicklung des Stellarators Wendelstein-7X legt die Vermutung nahe, dass die Nachfrage nach diesen Isotopen weiter steigen wird. [8]

Die Trennung von Molekülen aus Gasgemischen wird häufig mit Hilfe von Molekularsieben durchgeführt, die auf der unterschiedlichen Größe der zu trennenden Molekülen basiert. Isotope eines Elements besitzen aber dieselbe Größe, weshalb Molekularsiebe für deren Trennung nicht anwendbar sind. Heutzutage wird Deuterium mit Hilfe der kryogene Destillation bei 24 K oder mit dem Girdler-Sulfid (GS) Prozess gewonnen. Beide Methoden haben jedoch geringe Separationsfaktoren 1,5 für die kryogene Destillation und 2,3 für den GS Prozess bei einem gleichzeitig sehr hohem Energieverbrauch. Die Helium-3 Produktion ist eng mit der Tritiumproduktion der Nuklearwaffenindustrie verbunden, da Tritium zu Helium-3 zerfällt.

Die geringe Effizienz der konventionellen Isotopentrennung treibt die Forschung an, neue Verfahren zu entwickeln. Beenakker *et al.* schlugen im Jahr 1995 das Konzept des „Quantum Sieving“ vor, mit dem leichte Isotope in sehr engen Poren bei tiefen Temperaturen getrennt werden könnten. [10] Die Grundidee basiert auf der geringeren Nullpunktsenergie des schwereren Isotopes, welches dadurch leichter in die engen Poren eindrigen kann. Ein weiteres Konzept für die Isotopentrennung, das ebenfalls die verschiedenen Nullpunktsenergien nutzt, ist „Chemical Affinity Sieving". Der Unterschied der Nullpunktsenergien zwischen Isotopen kann durch die Adsorption an starken Adsorptionsplätzen vergrößert werden, was zu einer höheren Adsorptionsenthalpie des schweren Isotopes und letztendlich zur Trennung beider Isotope führt. [11]

Metallorganische Gerüststrukturen (engl. metal-organic frameworks, MOFs) bilden eine neue Klasse kristalliner poröser Materialien, die sich hervorragend für die Anwendung von „Quantum Sieving" oder „Chemical Affinity Sieving" eignen. Das Gerüst von MOFs ist modular aus anorganischen Bauelementen aufgebaut, welche über organische Linkermolekülen verknüpft werden. Aus der großen Variationsmöglichkeit der Module entstehen Materialien mit unterschiedlichen Porengrößen und vielfältigen Funktionalisierungen. Während für „Quantum Sieving" Materialien mit passenden Porengrößen benötigt werden, eignen sich für „Chemical Affinity Sieving" Materialien mit starken Adsorptionsplätzen, wie zum Beispiel ungesättigte Metallzentren.

An der Trennung von Wasserstoff- und Heliumisotopen wird aktuell auch in anderen Gruppen mit großem Interesse geforscht. Die Gruppe des Nobelpreisträgers Prof. Geim hat Wasserstoffisotopentrennung mit Hilfe von Graphen und Bornitrid Sieben bei Raumtemperatur durgeführt. [12] Sie erreichten einen Separationsfaktor von 10, jedoch wird mit diesen Sieben das Proton bevorzugt durchgelassen. Effizienter wäre es das seltene Deuteron zu extrahieren und dadurch nicht so viel Energie für den Transport der vielen Protonen aufzubringen. Heliumisotopentrennung wird aktuell ebenfalls mit Graphen untersucht. Theoretiker schufen ein Modell in dem zwei benachbarte Kohlenstoffringe entfernt wurden. [13,14] Die verbleibenden ungesättigten Verbindungen wurden mit Wasserstoff- und Stickstoffatome abgesättigt, sodass eine Potentialbarriere entstand, die Helium-3 bevorzugt durchlässt. Bei 10 K wurde ein Separationsfaktor von 19 errechnet, der bei 20 K auf 2 stark abfällt.

In dieser Arbeit wurde die Trennung von Wasserstoff- und Heliumisotope in metallorganischen Gerüststrukturen mit thermischer Desorptionsspektroskopie (TDS) experimentell untersucht. Die untersuchten Materialen wurden äquimolaren Isotopenmischungen bei bestimmten Temperaturen für eine gewisse Zeit ausgesetzt, und im Anschluss wurde ein Desorptionsspektrum aufgenommen. Aus diesem kann der Separationsfaktor, hier als Selektivität bezeichnet, direkt ermittelt werden, da die Flächen unter den TDS Maxima proportional zu der adsorbierten Menge sind.

Für diese Messungen wurden verschiedene metallorganische Gerüststrukturen mit geeigneten Porengrößen für das „Quantum Sieving“ von Wasserstoff- und Heliumisotopen ausgewählt. Passende Kandidaten bieten die IFP [15–17] und MFU-4 [18] Materialserien. „Chemical Affinity Sieving“ wird in dieser Arbeit an metallorganischen Gerüststrukturen mit ungesättigten Metallzentren untersucht. Ein Material mit sehr starken Kupfer Adsorptionsplätzen, Cu(I)-MFU-4*l* und drei Materialien mit ungesättigten Nickel Adsorptionsplätze, CPO-27 [20, 21], DUT-9 [22] und CFA-6 [23] wurden untersucht.

Messmethoden

Thermische Desorptionsspektroskopie (TDS)

Die thermische Desorptionsspektroskopie eignet sich sehr gut für die Charakterisierung der Wechselwirkung zwischen Gasen und Feststoffen. Diese Technik hat eine sehr hohe Sensitivität, sodass für Messungen nur wenige Milligram Probe ausreichen. In dieser Arbeit wurden Messungen bei tiefen Temperaturen in der hierfür speziell gebauten Anlage durchgeführt. Die Probe befindet sich vor der Messung im UHV und wird vor dem Beladen mit Gas auf eine bestimmte Expositionstemperatur zwischen 20 K und 120 K abgekühlt. Das Beladen mit Gas wurde abhängig von der untersuchen Probeneigenschaft auf verschiedene Arten durchgeführt. Isotope wurden in reiner Form (1 mbar bis 900 mbar), in einer äquimolaren Mischung (5 mbar bis 60 mbar) und nacheinander hinzugefügt (5 mbar oder 10 mbar). Die Expositionszeit varrierte zwischen 5 min und 180 min. Das nicht adsorbierte Gas wurde je nach Probe bei der Beladungstemperatur oder aber bei der niedrigsten erreichbaren Temperatur („Basistemperatur“, ungefähr 20 K) wieder abgepumpt. Das Aufheizen der Probe von der Basistemperatur bis zur Raumtemperatur erfolgte mit einer konstanten Aufheizrate und der Partialdruck der desorbierten Isotope wurde mit einem Quadrupolmassenspektrometer gemessen. Das TDS Spektrum zeigt an, bei welcher Temperatur die Isotope desorbieren. Die verschiedenen Maxima des Spektrums lassen auf verschiedene Adsorptionsplätze mit unterschiedlichen Aktivierungsenergien rückschließen. Mit Hilfe der Auswertemethode von Kissinger kann die Aktivierungsenergie bestimmter Adsorptionsplätze aus Spektren, die mit unterschiedlichen Heizraten aufgenommen wurden, ermittelt werden. Die Fläche unter dem TDS Spektrum ist proportional zur adsorbierten Menge. Für Wasserstoffisotope kann das Massenspektrometersignal mit Hilfe von Absorption in Metallhydriden kalibriert werden, sodass die Quantität der adsorbierten Menge bekannt ist. Das Beladen mit einer äquimolare Mischung macht es möglich aus der desorbierten Menge der einzelnen Isotope direkt das molare Adsorptionsverhältnis, die sogenannte Selektivität, zu ermitteln.

Inelastische Neutronenstreuung (INS)

Aufgrund des großen inkohärenten Streuquerschnittes des Wasserstoffkerns eignet sich INS sehr gut, für die Bestimmung seiner Wechselwirkung mit Materialien. Die Frequenz der Rotationsübergänge eines adsorbierten Wasserstoffmoleküls ändert sich im Vergleich zum freien Wasserstoff, wodurch Adsorptionsplätze mit underschiedlichen Aktivierungsenergien detektiert werden. Deuterium hat einen deutlich geringeren inkohärenten Streuquerschnitt als Wasserstoff, was dazu führt, dass mit Deuterium besetzte Adsorptionsplätze im INS Spektrum nicht sichtbar sind. Die INS Messungen dieser Arbeit wurden am Oak Ridge National Laboratory druchgeführt. [67] Cu(I)-MFU-4*l* wurde mit para-Wasserstoff beladen und INS Spektren wurden bei Temperaturen zwischen 5 K und 200 K aufgenommen. Zusätzlich zum Wasserstoff wurde für weitere Messreihen Deuterium hinzugefügt.

Raman-Spektroskopie

Die Vibrationsübergänge der homonuklearen Moleküle Wasserstoff und Deuterium können mit der optischen Messmethode der Raman-Spektroskopie untersucht werden. Dafür wurde die mit Gas beladene Probe mit einem Laser bestrahlt und die vom System zurück gestreute Strahlung wurde mit einem Spektrometer detektiert. Findet ein Übergang zwischen zwei verschiedene Energieniveaus statt, so gibt es eine Frequenzdifferenz zwischen eingestrahlter und zurück gestreuten Strahlung, die der Energieänderung des Moleküls entspricht. Durch Adsorption ändert sich die Vibrationsferquenz eines Moleküls im Vergleich zum freien Gasmolekül. Im Rahmen dieser Arbeit wurde die Wechselwirkung zwischen den Adsorptionsplätzen der metallorganischen Gerüststruktur Cu(I)-MFU-4*l* und den Wasserstoffisotopen bei Temperaturen zwischen 40 K und 200 K untersucht.

Ergebnisse und Diskussion

Adsorption und Trennung von Isotopen in engen Poren

Für die Wasserstoff- und Heliumisotopentrennung unter Ausnutzung des Effektes des „Quantum Sievings“ wurden spezielle metallorganische Gerüststrukturen mit passenden Porengrößen ausgewählt. „Quantum Sieving“ tritt auf, wenn die Bewegung der Atome oder Moleküle von Poren eingeschränkt wird, die nur unwesentlich größer sind als der kinetische Durchmesser der Atome oder Moleküle. Das schwerere Isotop hat eine niedrigere Nullpunktsenergie und diffundiert schneller in die engen Poren, was zur Trennung der Isotope führt.
Mit der Messmethode der thermischen Desorptionsspektroskopie wurden verschiedene Materialien der MFU-4 und IFP Serien für die mögliche Trennung von Wasserstoff- und Heliumisotopen untersucht. Diese Materialien haben Porenöffnungen mit Durchmessern im geeigneten Bereich für das „Quantum Sieving“ von Wasserstoff- und Heliumisotopen.

Bei den Materialien mit engen Poren, wirkt sich die Größe der Porenöffnung direkt auf das Desorptionsmaximum im TDS Spektrum aus. Kleinere Poren verringern die Diffusionsgeschwindigkeit aus dem Material und Maxima treten deshalb im TDS Spektrum bei höheren Temperaturen auf.
Die Porenöffnungen von IFP-4 (1,7 Å), IFP-7 (2,1 Å), Br-MFU-4 (2,33 Å) und Cl-MFU-4 (2,52 Å) sind viel kleiner als der kinetische Durchmesser der Wasserstoff- und Heliumisotope (2,89 Å, 2,60 Å [100, 104]). Die Flexibilität der die Öffnung bildenden Elemente ermöglicht trotzdem eine Gasaufnahme in den Materialien. Diese Gasaufnahme steigt mit längerer Expositionszeit an. Der sogenannte Porenöffnungseffekt (engl. gate opening effect) hängt von der Expositionstemperatur ab und reguliert das Adsorptionsverhalten dieser Materialien. In einer früheren Untersuchung wurde der Porenöffnungseffekt für Wasserstoffisotope in Cl-MFU-4 beobachtet. [102] Die Untersuchungen mit Helium zeigten jedoch, dass die Heliumaufnahme in Cl-MFU-4 bereits bei der niedrigsten Expositionstemperatur maximal ist und für höhere Expositionstemperaturen abfällt. Der nur sehr geringe Unterschied zwischen dem Durchmesser der Porenöffnung (2,52 Å) und dem kinetischen Durchmesser von Helium (2,60 Å) führt dazu, dass der Durchmesser der Pore aufgrund der Flexibilität bereits bei sehr niedrigen Temperaturen einen Wert erreicht, der die Heliumaufnahme ermöglicht. In dem untersuchten Temperaturbereich ist die Porenöffnung bereits in solchem Maße vergrößert, dass Heliumisotope am Eintritt ins Material kaum mehr gehin-

dert werden. Für Heliumisotope konnte der Porenöffnungseffekt in Br-MFU-4, welches mit 2,33 Å einen kleineren Porendurchmesser als Cl-MFU-4 hat, beobachtet werden. Die Temperaturabhängigkeit der Porenöffnung zeigte sich in TDS Spektren, aus denen sich schließen lässt, dass bei 30 K noch keine Heliumatome ins Material eindringen können, während die maximale Aufnahme bei 50 K erreicht ist.
Die Desorptionsenergie von Wasserstoffisotopen wurde mit Hilfe von TDS Messungen ermittelt und liegt für IFP-4 zwischen 6,6 kJ mol^{-1} und 7,0 kJ mol^{-1}. Die Werte stimmen gut mit dem durch Wassertsoffadsorptionsisothermen ermittelten Wert von 6,9 kJ mol^{-1} überein. [93] Die Desorptionsenergie von ^{4}He beträgt für IFP-4 5,1 kJ mol^{-1}. Im Vergleich mit der Desorptionsenergie von Wasserstoffisotopen ist dieser Wert etwas niedriger, was in Einklang mit dem Auftreten des Desorptionsmaximums von ^{4}He bei geringeren Temperaturen als bei Wasserstoff steht (103 K zu 115 K). Dieses Verhalten liegt am kleineren kinetischen Durchmesser von ^{4}He und dem daraus resultierenden erleichterten Vorankommen in diesen engen Kanälen.
Die Trennung von Wasserstoffisotopen wurde mit den Materialien der IFP Serie untersucht. Diese sind aus eindimensionalen Kanälen aufgebaut, die je nach Substituenten unterschiedlich große Porenöffnungen besitzen. Die Selektivität dieser Materialien ist kleiner als $S = 4$. Für IFP-4, dem Material mit der kleinsten Porenöffnung, wurde ein konstanter Wert von ungefähr $S = 2$ nach der Beladung mit verschiedene Drücken und bei verschiedenen Zeiten ermittelt. Der konstante Wert für die Selektivität ist auf die kanalähnliche Struktur mit kleinen Hohlräumen zwischen den Porenöffnungen zurückzuführen. Diese erlauben keinen Austausch der Isotope in den engen Kanälen und die Siebung geschieht nur an der äußeren Porenöffnung. In einer vorangegangenen Arbeit wurden Wasserstoffisotope in Cl-MFU-4 mit einer Selektivität von $S = 7,5$ getrennt. [102] Die MFU-4 Materialserie hat eine alternierende Porenstruktur bestehend aus großen und kleinen Poren, die durch eine kleine Porenöffnung miteinander verbunden sind. Das Volumen der großen Pore erlaubt eine Durchmischung der Wasserstoffisotope, weshalb an der kleinen Porenöffnung zwischen den großen und kleinen Poren das schwerer Isotop erneut bevorzugt durchgelassen wird. Da jeder Eindringvorgang in eine Nachbarpore zur weiteren Siebung beiträgt, führt dieser Vorgang zur Erhöhung der Selektivität. Dieses Ergebnis motivierte zur Heliumisotopentrennung in Cl-MFU-4 und Br-MFU-4. Die höchsten in dieser Arbeit ermittelten Selektivitäten betragen $S = 1,5$ für Br-MFU-4 und $S = 2,3$ für Cl-MFU-4. Die hohen Separationswerte, die für Wasserstoffisotope erreicht wurden, traten bei Heliumisotopen nicht auf.

Der kleinere Massenunterschied zwischen ^{3}He und ^{4}He verglichen mit H_2 und D_2 führt zu einen kleineren Unterschied in der Nullpunktsenergie. Dadurch verringert sich auch der Unterschied in der Diffusion zwischen den Heliumisotopen, was eine Verminderung der Separation durch „Quantum Sieving“ nach sich zieht.

Adsorption und Trennung von Isotopen an starken Adsorptionsplätzen

Metallorganische Gerüststrukturen mit ungesättigten Metallzentren wurden mit thermischer Desorptionsspektroskopie bei tiefen Temperaturen untersucht. Mit TDS können Adsorptionsplätze unterschiedlicher Bindungsstärke identifiziert werden, wobei die ungesättigten Metallzentren in den untersuchten Materialien die stärksten Adsorptionsplätze bilden. Deshalb werden die Maxima bei den höhsten Desorptionstemperaturen im Spektrum diesen Adsorptionsplätzen zugeschrieben. Starke Adsorptionsplätze sind äußerst wichtig für die Isotopentrennung mit „Chemical Affinity Sieving“, da sich der Unterschied in der Nullpunktsenergie beider Isotope für stark attraktive Potentiale vergrößert. Dadurch wird das schwerere Isotop mit der niedrigeren Nullpunktsenergie bevorzugt adsorbiert, was wiederum zu einer Trennung der Isotope führt.

Ein Material mit ungesättigten Metallzentren aus Kupfer, Cu(I)-MFU-4*l*, wurde tiefgehend in dieser Arbeit untersucht. In der Literatur veröffentlichte Messungen von Adsorptionsisothermen ergaben eine sehr große Adsorptionsenergie für H_2 von 32,3 $kJ\,mol^{-1}$ und aus Dichtefunktionaltheorie Rechnungen wurde ein Wert von 25 $kJ\,mol^{-1}$ ermittelt. [19] Ähnliche Werte wurden auch mit TDS in dieser Arbeit bestimmt: 24,8 $kJ\,mol^{-1}$ für H_2 und 33,2 $kJ\,mol^{-1}$ für D_2. Die Desorptionsmaxima dieses sehr starken Adsorptionsplatzes befinden sich im TDS Spektrum bei 166 K für H_2 und bei 181 K für D_2, was den höhsten Werten aller in dieser Arbeit untersuchten Materialien entspricht. Für starke Adsorptionspotentiale unterscheiden sich die Nullpunktenergien der Isotope sogar noch mehr, was dazu führt, dass die Desorptionsmaxima der Isotope bei unterschiedlichen Temperaturen auftreten. Das leichtere Isotop ist schwächer gebunden und desorbiert bei tieferen Temperaturen. Bis zur Messung von Cu(I)-MFU-4*l* hatte die desorbierte Menge an D_2 stets die desorbierte Menge an H_2 überstiegen. Bei Cu(I)-MFU-4*l* wurde zum ersten Mal beobachtet, dass am starken Adsorptionsplatz aus einer äquimolaren Isotopenmischung bei tiefen

Temperaturen (20 K, 30 K) mehr H_2 als D_2 adsorbiert wurde. Für Expositionstemperaturen zwischen 40 K und 77 K werden beide Isotope ungefähr im gleichen Maße adsorbiert und ab einer Temperatur oberhalb von 90 K wird fast nur D_2 adsorbiert. Dieser Austausch im Adsorptionsverhalten mit anwachsender Expositionstemperatur wurde mithilfe der Methode der 'stufenweisen Zufuhr' der Isotope untersucht. Für Expositionstemperaturen bis 77 K bleibt das zuerst hinzugefügte Isotop adsorbiert und es findet kein Austausch statt, wohingegen bei Expositionstemperaturen über 77 K das zuerst hinzugefügte H_2 durch das später ankommende D_2 ausgetauscht wird. Somit wurde gezeigt, dass ein Austausch des schneller diffundierenden H_2 durch D_2 am sehr starken Adsorptionsplatz nur bei höheren Expositionstemperaturen geschieht.
In Cu(I)-MFU-4*l* wurde die höhste Selektivität bei der Expositionstemperatur von 100 K erreicht. Für längere Beladezeiten steigt bei dieser Temperatur die Selektivität von $S = 10,8$ nach 5 min auf $S = 25,2$ nach 90 min, was zeigt, dass der Austauschprozess von H_2 durch D_2 stark zeitabhängig ist.
Mit inelastischer Neutronenstreuung (INS) wurde die Adsorption von H_2 in Cu(I)-MFU-4*l* ebenfalls untersucht. Bei einer geringen Beladung mit Wasserstoffgas wurde ein Doppelmaximum bei 4,9 meV und 6,5 meV beobachtet, das dem Rotationsübergang von $J = 0$ nach $J = 1$ des adsorbierten H_2 am ungesättigten Metallzentrum zugeordnet wird. Im Vergleich zum freien H_2 mit einer Bindungslänge von 0,75 Å [34] wird das an Cu(I) gebundene Wasserstoffmolekül zu 0,85 Å gestreckt. Durch Beladung mit einer höheren Menge an Wasserstoffgas wurden die starken Adsorptionsplätze vollständig belegt und neue Signale erscheinen im Spektrum, aufgrund der Adsorption des H_2 an schwächere Adsorptionsplätzen. Fügt man zusätzlich D_2 hinzu, so sinkt die Intensität der Maxima der starken Adsorptionsplätze im Spektrum, während die Intensität der Maxima der schwächeren Adsorptionsplätze gleich bleibt. Da D_2 im INS Spektrum nicht sichtbar ist, kann dies auf den Austausch des bereits adsorbierten H_2 an den starken Adsorptionsplätzen durch D_2 zurückgeführt werden. Dies wird durch TDS Messungen bestätigt.
Die Adsorption von Wasserstoffisotopen in Cu(I)-MFU-4*l* wurde auch mit Raman-Spektroskopie untersucht. Aufgrund der Adsorption ist die Vibration des Moleküls gestört und es treten zusätzlich zu den Signalen des ungestörten Moleküls weitere Signale im Spektrum auf. In dieser Arbeit wurden zur Q-Linie der freien H_2 und D_2 Moleküle rot-verschobene Signale beobachtet, die für beide Isotope um den gleichen prozentualen Anteil von der jeweiligen Q(0)-Linie verschoben sind. Dies ist ein Indiz dafür, dass die Isotope an den gleichen Adsorptionsplätzen gebunden sind. Das rot-verschobene Signal wird der Ad-

sorption an den schwachen Adsorptiosplätzen zugeschrieben. Die Adsorption von H_2 am Kupfer des Cu(I)-MFU-4*l* sollte zu einer weitaus größeren Verschiebung der Q-Linien führen. Rechnungen haben gezeigt, dass das Signal dieser Schwingung über einen weiten Frequenzbereich verschmiert ist und deshalb keine Beobachtung im Ramanspektrum möglich ist.

Es wurden drei metallorganische Gerüststrukturen mit Nickelatomen als ungesättigte Metallzentren untersucht: DUT-9, CPO-27 und CFA-6. Für DUT-9 und CPO-27 befindet sich das Desorptionsmaximum des starken Adsorptionsplatzes unterhalb und für CFA-6 oberhalb von 100 K. In CPO-27 und CFA-6 sind die Desorptionsmaxima des D_2 im Vergleich zu den Desorptionsmaxima des H_2 zu höheren Temperaturen verschoben. Für CPO-27 ergibt sich eine Temperaturverschiebung von 6 K und für CFA-6 von 14 K. Die höhere Nullpunktsenergie des leichteren Isotopes führt dazu, dass es schwächer gebunden ist und dadurch bei tieferen Temperaturen desorbiert. Die größere Temperaturdifferenz für CFA-6 (14 K) im Vergleich zu CPO-27 (6 K) tritt aufgrund des stärkeren Adsorptionsplatzes des CFA-6 auf. Dieser Zusammenhang ist in Übereinstimmung mit „Chemical Affinity Sieving", da sich die Nullpunktsenergien der Isotope für stärkere Adsorptionsplätze mehr unterscheiden.
An den starken Adsorptionsplätzen von CFA-6 wurde, wie schon bei Cu(I)-MFU-4*l*, ein Expositionstemperatur abhängiger Wechsel zwischen der bevorzugten Adsorption von H_2 und der von D_2 gemessen. Durch die 'stufenweise Zufuhr' der Isotope wurde bestätigt, dass es sich hierbei um den gleichen Vorgang handelt: bei niedrigen Expositionstemperaturen (30 K) konnte kein Austausch von bereits adsorbierten H_2 durch D_2 beobachtet werden, wogegen bei hohen Expositionstemperaturen (120 K) der Austausch auftrat. Für schwächere Adsorptionsplätze mit Desorptionsmaxima unter 100 K findet der Austausch bereits bei niedrigen Expositionstemperaturen statt. Das führt zu hohen Selektivitäten für DUT-9 $S = 35,8$ und CPO-27 $S = 18,5$.

Schlussfolgerung

Vergleicht man die in dieser Arbeit durchgeführten Untersuchungen von „Quantum Sieving" (QS) und „Chemical Affinity Sieving" (CAS), so zeigt CAS die besseren Selektivitäten und ist somit für die Isotopentrennung besser geeignet. Für QS muss die Apertur der Poren des benutzten MOFs etwas größer sein als der kinetische Durchmeser der zu trennenden Isotope. Sind die Poren zu groß,

so ist keine Trennung möglich, da beide Isotope gleich gut in das MOF hineingelangen können. Sind die Poren dagegen zu klein, kann keines der Isotope adsorbiert werden. Einige MOFs besitzen einen temperaturabhängigen Porendurchmesser, dessen Flexibilität einen negativen Einfluß auf die Separation mit QS haben kann. Im Fall von CAS hingegen, muss nur gewährleistet sein, dass die Isotope die starken Adsorptionsplätze erreichen können und die Porengröße ist somit nicht wesentlich.
Neben der starken Abhängigkeit vom Porendurchmesser, kommt es beim QS auch auf die Form der Poren an. Wenn Isotope in kanalähnlichen Poren nicht aneinader vorbei kommen, so kann die selektive Adsorption nur an der äußersten Apertur stattfinden. Befinden sich die Isotope in den Kanälen, kann kein weiteres Sieben im Material stattfinden. QS benötigt im Allgemeinen auch niedrigere Temperaturen als CAS, wo geeignete MOFs die Trennung bei Temperaturen weit über der Siedetemperatur von flüssigem Stickstoff ermöglichen (siehe Abbildung 7.1). Die besten in dieser Arbeit erreichten Resultate für die Isotopentrennung mit CAS wurden mit Cu(I)-MFU-4*l* erreicht, einem Material mit einem hohen Separationsfaktor bei Temperaturen weit oberhalb derer von flüssigem N_2. Die gefundenen Eigenschaften von Cu(I)-MFU-4*l* veranlassten zur experimentellen Nachstellung eines Temperaturwechsel Adsorptionsprozess zwischen 100 K und 220 K. Ein Wasserstoffisotopengemisch mit nur 5% D_2 wurde für 10 min dem Material bei 100 K hinzugefügt. Bereits nach dem ersten Adsorptions/Desorptionszyklus stieg in dem aus dem Cu(I)-MFU-4*l* desorbierten Gasmenge die D_2 Konzentration zu 81%. In einem weiteren Schritt wurde eine Konzentration von 93% D_2 erreicht. Der Unterschied in der Nullpunktsenergie der Isotope an starken Adsorptionsplätzen ermöglicht das schwerere Isotop aus einem Gemisch mit niedriger Konzentration einzufangen. Für Wasserstoffisotope ist dies eine ideale Voraussetzung, da das schwerere Isotop, D_2, eine sehr geringe natürliche Häufigkeit hat und deshalb effizient in wenigen Zyklen angereichert werden kann. Die Kosten für die Separation können so niedrig gehalten werden.
Die höchste Selektivität bei 100 K wurde in dieser Arbeit mit Cu(I)-MFU-4*l* erreicht, $S = 25,2$. Dieser hohe Separationsfaktor bei Temperaturen oberhalb der von flüssigem N_2 demonstriert eindrucksvoll die Möglichkeiten von CAS und wurde nach aktuellem Kenntnisstand bisher nicht übertroffen. Diese Arbeit ebnet den Weg für CAS, stellt aber auch die Frage nach den tatsächlichen Kosten der benötigten Materialien und damit der Umsetzbarkeit auf industriellen Größenordnungen. Die in dieser Arbeit benötigten Mengen an MOF überstiegen ein paar Milligramm nie, für Industrielle Anwendungen würden je-

doch zumindest Kilogramm benötigt. Die Chemikalien für die Herstellung von einem Kilogramm Cu(I)-MFU-4*l* kosten bei Einkauf in kleinen Mengen ungefähr 60000€. [135] Die Kosten für die benötigten Materialien werden bei der Herstellung in größerem Maßstab vermutlich stark verringert.
Die erzielten Ergebnisse zeigen das Potential von MOFs mit starken Adsorptionsplätzen für die effiziente Isotopentrennung. Mit einer Weiterentwicklung von MOFs ist zu rechnen, da allein in Deutschland über 40 Forschergruppen auf diesem Gebiet arbeiten. Zudem ist das Interesse der Industrie an MOFs für verschiedene Anwendungen, wie zum Beispiel Gasreinigung, Katalyse oder Gasspeicherung groß, was vermuten lässt, dass die Großproduktion von MOFs vorangetrieben wird.

Bibliography

[1] C. E. Housecroft and A. G. Sharpe, *Inorganic Chemistry, Third Edition*, ch. 3.10. Pearson Prentice Hall, 2008.

[2] R. E. Stanley and A. A. Moghissi, *Noble Gases*. U.S. Environmental Protection Agency, 1975.

[3] M. Baker and T. W. Crane, *Passive Nondestructive Assay of Nuclear Materials*, ch. 13. Neutron Detectors. Los Alamos National Laboratory, 1991.

[4] H. E. Hall, P. J. Ford, and K. Thompson, "A helium-3 dilution refrigerator," *Cryogenics*, vol. 6, no. 2, pp. 80–8, 1966.

[5] R. Radebaug and J. Siegwart, "Dilution refrigerator technology," *Cryogenics*, vol. 11, no. 5, pp. 368–84, 1971.

[6] L. J. Wittenberg, J. F. Santarius, and G. L. Kulcinski, "Lunar Source of He-3 for Commercial Fusion Power," *Fusion Technology*, vol. 10, no. 2, pp. 167–178, 1986.

[7] W. Z. Fa and Y. Q. Jin, "Quantitative estimation of helium-3 spatial distribution in the lunar regolith layer," *Icarus*, vol. 190, no. 1, pp. 15–23, 2007.

[8] T. S. Pedersen, T. Andreeva, H. S. Bosch, S. Bozhenkov, F. Effenberg, M. Endler, Y. Feng, D. A. Gates, J. Geiger, D. Hartmann, H. Holbe, M. Jakubowski, R. Konig, H. P. Laqua, S. Lazerson, M. Otte, M. Preynas, O. Schmitz, T. Stange, Y. Turkin, and W.-X. Team, "Plans for the first plasma operation of Wendelstein 7-X," *Nuclear Fusion*, vol. 55, no. 12, 2015.

[9] H. K. Rae, "Separation of hydrogen isotopes," *American Chemical Society*, vol. 68, pp. 1–26, 1978.

[10] J. J. M. Beenakker, V. D. Borman, and S. Y. Krylov, "Molecular-transport in subnanometer pores - zero-point energy, reduced dimensionality and quantum sieving," *Chemical Physics Letters*, vol. 232, no. 4, pp. 379–382, 1995.

[11] I. Savchenko, A. Mavrandonakis, T. Heine, H. Oh, J. Teufel, and M. Hirscher, "Hydrogen isotope separation in metal-organic frameworks: Kinetic or chemical affinity quantum-sieving?," *Microporous and Mesoporous Materials*, vol. 216, pp. 133–137, 2015.

[12] M. Lozada-Hidalgo, S. Hu, O. Marshall, A. Mishchenko, A. N. Grigorenko, R. A. W. Dryfe, B. Radha, I. V. Grigorieva, and A. K. Geim, "Sieving hydrogen isotopes through two-dimensional crystals," *Science*, vol. 351, no. 6268, pp. 68–70, 2016.

[13] A. W. Hauser and P. Schwerdtfeger, "Nanoporous Graphene Membranes for Efficient He-3/He-4 Separation," *Journal of Physical Chemistry Letters*, vol. 3, no. 2, pp. 209–213, 2012.

[14] A. W. Hauser, J. Schrier, and P. Schwerdtfeger, "Helium tunneling through nitrogen-functionalized graphene pores: Pressure- and temperature-driven approaches to isotope separation," *Journal of Physical Chemistry C*, vol. 116, no. 19, pp. 10819–10827, 2012.

[15] F. Debatin, K. Behrens, J. Weber, I. A. Baburin, A. Thomas, J. Schmidt, I. Senkovska, S. Kaskel, A. Kelling, N. Hedin, Z. Bacsik, S. Leoni, G. Seifert, C. Jager, C. Gunter, U. Schilde, A. Friedrich, and H. J. Holdt, "An isoreticular family of microporous metal-organic frameworks based on zinc and 2-substituted imidazolate-4-amide-5-imidate: Syntheses, structures and properties," *Chemistry-a European Journal*, vol. 18, no. 37, pp. 11630–11640, 2012.

[16] F. Debatin, A. Thomas, A. Kelling, N. Hedin, Z. Bacsik, I. Senkovska, S. Kaskel, M. Junginger, H. Muller, U. Schilde, C. Jager, A. Friedrich, and H. J. Holdt, "In Situ Synthesis of an Imidazolate-4-amide-5-imidate Ligand and Formation of a Microporous Zinc-Organic Framework with H_2-and CO_2-Storage Ability," *Angewandte Chemie-International Edition*, vol. 49, no. 7, pp. 1258–1262, 2010.

[17] S. S. Mondal, A. Bhunia, I. A. Baburin, C. Jager, A. Kelling, U. Schilde, G. Seifert, C. Janiak, and H. J. Holdt, "Gate effects in a hexagonal

zinc-imidazolate-4-amide-5-imidate framework with flexible methoxy substituents and co2 selectivity," *Chemical Communications*, vol. 49, no. 69, pp. 7599–7601, 2013.

[18] S. Biswas, M. Grzywa, H. P. Nayek, S. Dehnen, I. Senkovska, S. Kaskel, and D. Volkmer, "A cubic coordination framework constructed from benzobistriazolate ligands and zinc ions having selective gas sorption properties," *Dalton Trans*, no. 33, pp. 6487–95, 2009.

[19] D. Denysenko, M. Grzywa, J. Jelic, K. Reuter, and D. Volkmer, "Scorpionate-Type Coordination in MFU-4l Metal-Organic Frameworks: Small-Molecule Binding and Activation upon the Thermally Activated Formation of Open Metal Sites," *Angewandte Chemie-International Edition*, vol. 53, no. 23, pp. 5832–5836, 2014.

[20] P. D. C. Dietzel, Y. Morita, R. Blom, and H. Fjellvag, "An in situ high-temperature single-crystal investigation of a dehydrated metal-organic framework compound and field-induced magnetization of one-dimensional metaloxygen chains," *Angewandte Chemie-International Edition*, vol. 44, no. 39, pp. 6354–6358, 2005.

[21] P. D. C. Dietzel, R. E. Johnsen, R. Blom, and H. Fjellvag, "Structural changes and coordinatively unsaturated metal atoms on dehydration of honeycomb analogous microporous metal-organic frameworks," *Chemistry-a European Journal*, vol. 14, no. 8, pp. 2389–2397, 2008.

[22] K. Gedrich, I. Senkovska, N. Klein, U. Stoeck, A. Henschel, M. R. Lohe, I. A. Baburin, U. Müller, and S. Kaskel, "A highly porous metal-organic framework with open nickel sites," *Angewandte Chemie-International Edition*, vol. 49, no. 45, pp. 8489–8492, 2010.

[23] C. Pettinari, A. Tabacaru, I. Boldog, K. V. Domasevitch, S. Galli, and N. Masciocchi, "Novel coordination frameworks incorporating the 4,4 '-bipyrazolyl ditopic ligand," *Inorganic Chemistry*, vol. 51, no. 9, pp. 5235–5245, 2012.

[24] W. J. Thomas and B. Crittenden, *Adsorption Technology & Design*. Elsevier Science & Technology, 1998.

[25] J. E. Lennard-Jones, "Cohesion," *Proceedings of the Physical Society*, vol. 43, pp. 461–482, 1931.

[26] F. London, "On the theory and systematic of molecular forces," *Zeitschrift Fur Physik*, vol. 63, no. 3-4, pp. 245–279, 1930.

[27] D. H. Everett and J. C. Powl, "Adsorption in slit-like and cylindrical micropores in henrys law region - model for microporosity of carbons," *Journal of the Chemical Society-Faraday Transactions I*, vol. 72, pp. 619–636, 1976.

[28] J. A. R. Navarro, E. Barea, J. M. Salas, N. Masciocchi, S. Galli, A. Sironi, C. O. Ania, and J. B. Parra, "H_2, N_2, CO, and CO_2 sorption properties of a series of robust sodalite-type microporous coordination polymers," *Inorganic Chemistry*, vol. 45, no. 6, pp. 2397–2399, 2006.

[29] X. Lin, J. H. Jia, P. Hubberstey, M. Schroder, and N. R. Champness, "Hydrogen storage in metal-organic frameworks," *Crystengcomm*, vol. 9, no. 6, pp. 438–448, 2007.

[30] D. J. Collins and H. C. Zhou, "Hydrogen storage in metal-organic frameworks," *Journal of Materials Chemistry*, vol. 17, no. 30, pp. 3154–3160, 2007.

[31] B. Panella, K. Hones, U. Muller, N. Trukhan, M. Schubert, H. Putter, and M. Hirscher, "Desorption studies of hydrogen in metal-organic frameworks," *Angewandte Chemie-International Edition*, vol. 47, no. 11, pp. 2138–2142, 2008.

[32] G. J. Kubas, R. R. Ryan, B. I. Swanson, P. J. Vergamini, and H. J. Wasserman, "Characterization of the 1st Examples of Isolable Molecular-Hydrogen Complexes, Mo(Co)3(Pcy3)2(H2), W(Co)3(Pcy3)2(H2),Mo(Co)3 (Pi-Pr3)2(H2),W(Co)3(Pi-Pr3)2(H2) - Evidence for a Side-on Bonded H-2 Ligand," *Journal of the American Chemical Society*, vol. 106, no. 2, pp. 451–452, 1984.

[33] G. J. Kubas, "Fundamentals of H-2 binding and reactivity on transition metals underlying hydrogenase function and H-2 production and storage," *Chemical Reviews*, vol. 107, no. 10, pp. 4152–4205, 2007.

[34] P. Mitchell, S. Parker, A. Ramirez-Cuesta, and J. Tomkinson, *Vibrational spectroscopy with neutrons with applications in chemistry, biology, material science and catalysis*. Singapore: World Scientific, 2005.

[35] J. Eckert, "Dynamics of the molecular-hydrogen ligand in metal-complexes," *Spectrochimica Acta Part a-Molecular and Biomolecular Spectroscopy*, vol. 48, no. 3, pp. 363–378, 1992.

[36] M. Ragheb, *Isotopic Separation and Enrichment. Nuclear, Plasma and Radiation Science: Inventing the Future.* University of Illinois at Urbana-Champaign, USA, 2014.

[37] E. Riedel and C. Janiak, *Anorganische Chemie*, ch. 4, table 4.2. De Gruyter, 2011.

[38] R. T. Kouzes, *The ^{3}He supply problem.* Pacific Northwest National Laboratory, Richland, WA (Technical Report PNNL-18388), 2009.

[39] D. A. Shea and D. Morgan, *The Helium-3 Shortage: Supply, Demand, and Options for Congress.* CRS Report for Congress, Congressional Research Service, 2010.

[40] K. Nakamura and H. Thomas, "Quantum billiard in a magnetic-field - chaos and diamagnetism," *Physical Review Letters*, vol. 61, no. 3, pp. 247–250, 1988.

[41] T. X. Nguyen, H. Jobic, and S. K. Bhatia, "Microscopic observation of kinetic molecular sieving of hydrogen isotopes in a nanoporous material," *Physical Review Letters*, vol. 105, no. 8, 2010.

[42] Y. Jiao, A. J. Du, M. Hankel, and S. C. Smith, "Modelling carbon membranes for gas and isotope separation," *Physical Chemistry Chemical Physics*, vol. 15, no. 14, pp. 4832–4843, 2013.

[43] T. E. Madey and J. T. Yates, *Vibrational Spectroscopy of Molecules on Surfaces.* New York: Springer Science+Business Media, LLC, 1987.

[44] H. Oh, I. Savchenko, A. Mavrandonakis, T. Heine, and M. Hirscher, "Highly effective hydrogen isotope separation in nanoporous metal-organic frameworks with open metal sites: Direct measurement and theoretical analysis," *ACS Nano*, vol. 8, no. 1, pp. 761–770, 2013.

[45] S. Bhatia, J. Beltramini, and D. Do, *Temperature Programmed Analysis and Its Applications in Catalytic Systems.* Elsevier, 1990.

[46] P. Redhead, "Thermal desorption of gases," *Vacuum*, vol. 12, no. 4, pp. 203 – 211, 1962.

[47] G. Carter, "Thermal resolution of desorption energy spectra," *Vacuum*, vol. 12, no. 5, pp. 245–254, 1962.

[48] M. Polanyi and E. Wigner, "Concerning the interference of natural oscillation as a reason for energy variations and chemical transformation.," *Zeitschrift Fur Physikalische Chemie–Stochiometrie Und Verwandtschaftslehre*, vol. 139, pp. 439–452, 1928.

[49] H. E. Kissinger, "Reaction kinetics in differential thermal analysis," *Analytical Chemistry*, vol. 29, no. 11, pp. 1702–1706, 1957.

[50] A. M. de Jong and J. W. Niemantsverdriet, "Thermal desorption analysis: Comparative test of ten commonly applied procedures," *Surface Science*, vol. 233, no. 3, pp. 355–365, 1990.

[51] F. M. Lord and J. Kittelberger, "Determination of activation-energies in thermal desorption experiments," *Surface Science*, vol. 43, no. 1, pp. 173–182, 1974.

[52] J. L. Falconer and R. J. Madix, "Flash desorption activation-energies - dcooh decomposition and co desorption from ni(110)," *Surface Science*, vol. 48, no. 2, pp. 393–405, 1975.

[53] J. L. Falconer and R. J. Madix, "Flash Desorption Activation-Energies - Dcooh Decomposition and Co Desorption from Ni(110)," *Surface Science*, vol. 48, no. 2, pp. 393–405, 1975.

[54] D. P. Woodruff and T. A. Delchar, *Modern Techniques of Surface Science*. Cambrdige Solid State Science Series, 1994.

[55] H. Haken and C. Wolf, *Molecular Physics and Elements of Quantum Chemistry*, vol. 5. Berlin Heidelberg: Springer-Verlag, 2005.

[56] P. C. Souers, *Hydrogen Properties for Fusion Energy*. Berkeley, Los Angeles, London: University of California Press, 1986.

[57] I. F. Silvera, "The solid molecular hydrogens in the condensed phase - fundamentals and static properties," *Reviews of Modern Physics*, vol. 52, no. 2, pp. 393–452, 1980.

[58] S. Sherif, D. Y. Goswami, E. K. Stefanakos, and A. Steinfeld, *Handbook of Hydrogen Energy.* Boca Raton: CRC Press, Taylor & Francis Group, 2014.

[59] V. F. Sears, "Neutron scattering lengths and cross sections of the elements and their isotopes," vol. 3, pp. 29–37, 1992.

[60] A. J. Ramirez-Cuesta, P. C. H. Mitchell, and S. F. Parker, "An inelastic neutron scattering study of the interaction of dihydrogen with the cobalt site of a cobalt aluminophosphate catalyst - two-dimensional quantum rotation of adsorbed dihydrogen," *Journal of Molecular Catalysis a-Chemical*, vol. 167, no. 1-2, pp. 217–224, 2001.

[61] B. Ludescher, *Construction plan for the thermal desorption spectroscopy apparatus.* Max Planck Insitute for Intelligent Systems, Stuttgart, 2011.

[62] M. Yoshihara and R. B. Mclellan, "Palladium Cerium Hydrogen Solid-Solutions 1. Thermodynamic Properties," *Acta Metallurgica*, vol. 36, no. 12, pp. 3217–3224, 1988.

[63] M. Yoshihara and R. B. Mclellan, "Palladium Cerium Hydrogen Solid-Solutions 2. Kinetic-Properties," *Acta Metallurgica*, vol. 36, no. 12, pp. 3225–3229, 1988.

[64] M. Ege and H. Kronmüller, "Mechanical relaxation processes of hydrogen in dilute palladium rare earth alloys," *Journal of Alloys and Compounds*, vol. 261, no. 1-2, pp. 231–241, 1997.

[65] F. von Zeppelin, M. Haluska, and M. Hirscher, "Thermal desorption spectroscopy as a quantitative tool to determine the hydrogen content in solids," *Thermochimica Acta*, vol. 404, no. 1-2, pp. 251–258, 2003.

[66] S. R. Challa, D. S. Sholl, and J. K. Johnson, "Light isotope separation in carbon nanotubes through quantum molecular sieving," *Physical Review B*, vol. 63, no. 24, 2001.

[67] "Oak ridge national laboratory neutron sciences, beamline 16b." `https://neutrons.ornl.gov/vision`. Accessed: 2016-02-29.

[68] B. Panella and M. Hirscher, "Raman studies of hydrogen adsorbed on nanostructured porous materials," *Physical Chemistry Chemical Physics*, vol. 10, no. 20, pp. 2910–2917, 2008.

[69] S. Kaskel, F. Schüth, K. S. W. Sing, and J. Weitkamp, *Porous Metal-Organic Frameworks*, vol. 2, ch. 4.3, pp. 1190–1249. 69469 Weinheim: WILEY-VCH Verlag GmbH, 2002.

[70] H. C. Zhou, J. R. Long, and O. M. Yaghi, "Introduction to metal-organic frameworks," *Chemical Reviews*, vol. 112, no. 2, pp. 673–674, 2012.

[71] H. C. Zhou and S. Kitagawa, "Metal-Organic Frameworks (MOFs)," *Chemical Society Reviews*, vol. 43, no. 16, pp. 5415–5418, 2014.

[72] B. F. Hoskins and R. Robson, "Infinite polymeric frameworks consisting of 3 dimensionally linked rod-like segments," *Journal of the American Chemical Society*, vol. 111, no. 15, pp. 5962–5964, 1989.

[73] B. F. Abrahams, B. F. Hoskins, J. P. Liu, and R. Robson, "The Archetype for a New Class of Simple Extended 3d Honeycomb Frameworks - the Synthesis and X-Ray Crystal-Structures of Cd(Cn)5/3(Oh)1/3.1/3(C6h12n4), Cd(Cn)2.1/3(C6h12n4), and Cd(Cn)2.2/3h2o.Tbuoh (C6h12n4 = Hexamethylenetetramine) Revealing 2 Topologically Equivalent but Geometrically Different Frameworks," *Journal of the American Chemical Society*, vol. 113, no. 8, pp. 3045–3051, 1991.

[74] B. F. Hoskins and R. Robson, "Design and Construction of a New Class of Scaffolding-Like Materials Comprising Infinite Polymeric Frameworks of 3-D-Linked Molecular Rods - a Reappraisal of the Zn(Cn)2 and Cd(Cn)2 Structures and the Synthesis and Structure of the Diamond-Related Frameworks [N(Ch3)4][Cuiznii(Cn)4] and Cui[4,4',4",4"'-Tetracyanotetraphenylmethane]Bf4.Xc6h5no2," *Journal of the American Chemical Society*, vol. 112, no. 4, pp. 1546–1554, 1990.

[75] O. M. Yaghi, G. M. Li, and H. L. Li, "Selective binding and removal of guests in a microporous metal-organic framework," *Nature*, vol. 378, no. 6558, pp. 703–706, 1995.

[76] H. Li, M. Eddaoudi, M. O'Keeffe, and O. M. Yaghi, "Design and synthesis of an exceptionally stable and highly porous metal-organic framework," *Nature*, vol. 402, no. 6759, pp. 276–279, 1999.

[77] M. Eddaoudi, D. B. Moler, H. L. Li, B. L. Chen, T. M. Reineke, M. O'Keeffe, and O. M. Yaghi, "Modular chemistry: Secondary building

units as a basis for the design of highly porous and robust metal-organic carboxylate frameworks," *Accounts of Chemical Research*, vol. 34, no. 4, pp. 319–330, 2001.

[78] A. Centrone, D. Y. Siberio-Perez, A. R. Millward, O. M. Yaghi, A. J. Matzger, and G. Zerbi, "Raman spectra of hydrogen and deuterium adsorbed on a metal-organic framework," *Chemical Physics Letters*, vol. 411, no. 4-6, pp. 516–519, 2005.

[79] M. Eddaoudi, J. Kim, N. Rosi, D. Vodak, J. Wachter, M. O'Keeffe, and O. M. Yaghi, "Systematic design of pore size and functionality in isoreticular mofs and their application in methane storage," *Science*, vol. 295, no. 5554, pp. 469–472, 2002.

[80] N. Stock and S. Biswas, "Synthesis of Metal-Organic Frameworks (MOFs): Routes to Various MOF Topologies, Morphologies, and Composites," *Chemical Reviews*, vol. 112, no. 2, pp. 933–969, 2012.

[81] M. Herrmann, P. B. Kempa, H. Fietzek, T. Altenburg, A. Polyzoidis, C. G. Piscopo, and S. Löbbecke, "Characterization ol Metal-Organic Frameworks Using X-ray Diffraction," *Chem. Ing. Tech.*, vol. 88, pp. 1–5, 2016.

[82] M. Kurmoo, "Magnetic metal-organic frameworks," *Chemical Society Reviews*, vol. 38, no. 5, pp. 1353–1379, 2009.

[83] G. Givaja, P. Amo-Ochoa, C. J. Gomez-Garcia, and F. Zamora, "Electrical conductive coordination polymers," *Chemical Society Reviews*, vol. 41, no. 1, pp. 115–147, 2012.

[84] M. V. Lucky, S. Sivakumar, M. L. P. Reddy, A. K. Paul, and S. Natarajan, "Lanthanide Luminescent Coordination Polymer Constructed from Unsymmetrical Dinuclear Building Blocks Based on 4-((1H-Benzo[d]imidazol-1-yl)methyl) benzoic Acid," *Crystal Growth & Design*, vol. 11, no. 3, pp. 857–864, 2011.

[85] I. Eryazici, O. K. Farha, O. C. Compton, C. Stern, J. T. Hupp, and S. T. Nguyen, "Luminescent infinite coordination polymer materials from metal-terpyridine ligation," *Dalton Transactions*, vol. 40, no. 36, pp. 9189–9193, 2011.

[86] B. Zhao, X. Y. Chen, P. Cheng, D. Z. Liao, S. P. Yan, and Z. H. Jiang, "Coordination polymers containing 1D channels as selective luminescent probes," *Journal of the American Chemical Society*, vol. 126, no. 47, pp. 15394–15395, 2004.

[87] S. Cao, "Metal-organic frameworks: A new class of crystalline porous materials," *Johnson Matthey Technol. Rev.*, vol. 59, no. 2, pp. 123–125, 2015.

[88] U. Müller, M. Schubert, F. Teich, H. Puetter, K. Schierle-Arndt, and J. Pastre, "Metal-organic frameworks - prospective industrial applications," *Journal of Materials Chemistry*, vol. 16, no. 7, pp. 626–636, 2006.

[89] G. Ferey, "Hybrid porous solids: past, present, future," *Chemical Society Reviews*, vol. 37, no. 1, pp. 191–214, 2008.

[90] J. W. Liu, L. F. Chen, H. Cui, J. Y. Zhang, L. Zhang, and C. Y. Su, "Applications of metal-organic frameworks in heterogeneous supramolecular catalysis," *Chemical Society Reviews*, vol. 43, no. 16, pp. 6011–6061, 2014.

[91] B. Panella and M. Hirscher, *Handbook of Hydrogen Storage*, ch. 2.5. Edited by M. Hirscher, Wiley-VCH Weinheim, 2010.

[92] J. W. Ren, H. W. Langmi, B. C. North, and M. Mathe, "Review on processing of metal-organic framework (MOF) materials towards system integration for hydrogen storage," *International Journal of Energy Research*, vol. 39, no. 5, pp. 607–620, 2015.

[93] D. Denysenko, *Isosteric heats of hydrogen adsorption in IFP frameworks-a handout.* Institute of Physics, University of Augsburg, Augsburg, 2015.

[94] T. Duren, F. Millange, G. Ferey, K. S. Walton, and R. Q. Snurr, "Calculating geometric surface areas as a characterization tool for metal-organic frameworks," *Journal of Physical Chemistry C*, vol. 111, no. 42, pp. 15350–15356, 2007.

[95] D. Denysenko, M. Grzywa, M. Tonigold, B. Streppel, I. Krkljus, M. Hirscher, E. Mugnaioli, U. Kolb, J. Hanss, and D. Volkmer, "Elucidating Gating Effects for Hydrogen Sorption in MFU-4-Type Triazolate-Based Metal-Organic Frameworks Featuring Different Pore Sizes," *Chemistry-a European Journal*, vol. 17, no. 6, pp. 1837–1848, 2011.

[96] "Dut-9 data sheet." `http://www.chm.tu-dresden.de/ac1/materials_center/pdf/adsorbentien/dut9.pdf`. Accessed: 2016-03-29.

[97] T. G. Glover, G. W. Peterson, B. J. Schindler, D. Britt, and O. Yaghi, "MOF-74 building unit has a direct impact on toxic gas adsorption," *Chemical Engineering Science*, vol. 66, no. 2, pp. 163–170, 2011.

[98] P. D. C. Dietzel, B. Panella, M. Hirscher, R. Blom, and H. Fjellvag, "Hydrogen adsorption in a nickel based coordination polymer with open metal sites in the cylindrical cavities of the desolvated framework," *Chemical Communications*, no. 9, pp. 959–961, 2006.

[99] S. Spirkl, M. Grzywa, C. S. Zehe, J. Senker, S. Demeshko, F. Meyer, S. Riegg, and D. Volkmer, "Fe/Ga-CFA-6-metal organic frameworks featuring trivalent metal centers and the 4,4 '-bipyrazolyl ligand," *Crystengcomm*, vol. 17, no. 2, pp. 313–322, 2015.

[100] A. F. Ismail, K. Khulbe, and T. Matsuura, *Gas Separation Membranes: Polymeric and Inorganic*. Springer, 2015.

[101] H. Oh, K. S. Park, S. B. Kalidindi, R. A. Fischer, and M. Hirscher, "Quantum cryo-sieving for hydrogen isotope separation in microporous frameworks: an experimental study on the correlation between effective quantum sieving and pore size," *Journal of Materials Chemistry A*, vol. 1, no. 10, pp. 3244–3248, 2013.

[102] J. Teufel, H. Oh, M. Hirscher, M. Wahiduzzaman, L. Zhechkov, A. Kuc, T. Heine, D. Denysenko, and D. Volkmer, "MFU-4 - A Metal-Organic Framework for Highly Effective H-2/D-2 Separation," *Advanced Materials*, vol. 25, no. 4, pp. 635–639, 2013.

[103] K. S. W. Sing, D. H. Everett, R. A. W. Haul, L. Moscou, R. A. Pierotti, J. Rouquerol, and T. Siemieniewska, "Reporting physisorption data for gas/solid systems with special reference to the determination of surface area and porosity," *Pure Appl. Chem.*, vol. 57, pp. 603–619, 1985.

[104] Y. Matteucci, S. andYampolskii, B. Freeman, and I. Pinnau, *Transport of gases and vapors in glassy and rubbery polymers in Materials Science of Membranes for Gas and Vapor Separation*. John Wiley and Sons, 2006.

[105] M. Schlichtenmayer and M. Hirscher, "Nanosponges for hydrogen storage," *Journal of Materials Chemistry*, vol. 22, no. 20, pp. 10134–10143, 2012.

[106] J. S. Teufel, *Doctoral thesis: Experimental investigation of H_2/ D_2 isotope separation by cryo-adsorption in metal-organic frameworks.* Universtät Stuttgart, Germany, 2012.

[107] D. Martins, I. Catarino, D. Lopes, I. Esteves, J. Mota, and G. Bonfait, "Low Temperature Adsorption Versus Pore Size in Activated Carbons," *Cryocoolers 16, edited by S.D. Miller and R.G. Ross, Jr.*, pp. 567–573, 2011.

[108] N. Wada, K. Ishioh, and T. Watanabe, "Dependence of Helium Adsorption Energy on the Amount Adsorbed in Na-Y Zeolitic Pores," *Journal of The Physical Society of Japan*, vol. 61, no. 3, pp. 931–936, 1992.

[109] I. Krkljus, *Doctoral thesis: Correlation between the Microstructure of Porous Materials and the Adsorption Properties of H_2 and D_2.* Universität Stuttgart, Germany, 2011.

[110] *Personal communication with D. Denysenko.* Max Planck Institute for Intelligent Systems, Stuttgart, Germany, 2014.

[111] I. Weinrauch, I. Savchenko, D. Denysenko, S. M. Souliou, H. H. Kim, M. Le Tacon, L. L. Daemen, Y. Cheng, A. Mavrandonakis, A. J. Ramirez-Cuesta, D. Volkmer, M. Hirscher, and T. Heine, "D_2 and T_2 capture in the metal-organic framework Cu(I)-MFU-4l," *Submitted to Nature Communications*, 2016.

[112] A. P. Smith, R. Benedek, F. R. Trouw, M. Minkoff, and L. H. Yang, "Quasi-two-dimensional quantum states of H-2 in stage-2 Rb-intercalated graphite," *Physical Review B*, vol. 53, no. 15, pp. 10187–10199, 1996.

[113] J. A. Young and J. U. Koppel, "Slow neutron scattering by molecular hydrogen and deuterium," *Physical Review*, vol. 135, no. 3A, pp. A603–A611, 1964.

[114] *Personal communication with A. J. Ramirez-Cuesta.* Oak Ridge National Laboratory, Tennessee, USA, 2014.

[115] R. B. E. Down, A. J. Ramirez-Cuesta, R. A. Major, J. Keeping, S. Rudic, and O. Kirichek, "Cryogenic sample environment on tosca," *Dynamics of Molecules and Materials-Ii*, vol. 554, 2014.

[116] B. P. Stoicheff, "High Resolution Raman Spectroscopy of Gases .9. Spectra of H_2, HD, and D_2," *Canadian Journal of Physics*, vol. 35, no. 6, pp. 730–741, 1957.

[117] J. Fiutak and J. Van Kranendonk, "IMPACT THEORY OF RAMAN LINE BROADENING," *Canadian Journal of Physics*, vol. 40, pp. 1085–1100, 1962.

[118] J. Fiutak, "THE MULTIPOLE EXPANSION IN QUANTUM THEORY," *Canadian Journal of Physics*, vol. 41, pp. 12–20, 1963.

[119] A. D. May, H. L. Welsh, V. Degen, and J. C. Stryland, "Raman Effect in Gaseous Hydrogen at High Pressures," *Canadian Journal of Physics*, vol. 39, no. 12, pp. 1769–1783, 1961.

[120] A. D. May, G. Varghese, H. L. Welsh, and J. C. Stryland, "Vibrational Frequency Perturbations in Raman Spectrum of Compressed Gaseous Hydrogen," *Canadian Journal of Physics*, vol. 42, no. 6, pp. 1058–1069, 1964.

[121] E. C. Looi, J. C. Stryland, and H. L. Welsh, "Pressure Shifts in Vibrational Raman-Spectra of Hydrogen and Deuterium, 315K-85K," *Canadian Journal of Physics*, vol. 56, no. 8, pp. 1102–1108, 1978.

[122] A. Centrone, L. Brambilla, T. Renouard, L. Gherghel, C. Mathis, K. Mullen, and G. Zerbi, "Structure of new carbonaceous materials: The role of vibrational spectroscopy," *Carbon*, vol. 43, no. 8, pp. 1593–1609, 2005.

[123] S. S. Bhatnagar, H. L. Welsh, and E. J. Allin, "Raman Spectra of Liquid and Solid H_2, D_2, and HD at High Resolution," *Canadian Journal of Physics*, vol. 40, no. 1, pp. 9–23, 1962.

[124] A. D. Buckingham, "Solvent effects in vibrational spectroscopy," *Transactions of the Faraday Society*, vol. 56, no. 6, pp. 753–760, 1960.

[125] S. A. FitzGerald, J. N. Nelson, E. Gilmour, and J. L. C. Rowsell, "Infrared overtone spectroscopy of adsorbed hydrogen in mof-5," *Journal of Molecular Spectroscopy*, vol. 307, pp. 20–26, 2015.

[126] K. A. Williams, B. K. Pradhan, P. C. Eklund, M. K. Kostov, and M. W. Cole, "Raman spectroscopic investigation of H_2, HD, and D_2 physisorption on ropes of single-walled, carbon nanotubes," *Physical Review Letters*, vol. 88, no. 16, 2002.

[127] J. G. Vitillo, L. Regli, S. Chavan, G. Ricchiardi, G. Spoto, P. D. C. Dietzel, S. Bordiga, and A. Zecchina, "Role of exposed metal sites in hydrogen storage in MOFs," *Journal of the American Chemical Society*, vol. 130, no. 26, pp. 8386–8396, 2008.

[128] S. A. FitzGerald, J. Hopkins, B. Burkholder, M. Friedman, and J. L. C. Rowsell, "Quantum dynamics of adsorbed normal- and para- H_2, HD, and D_2 in the microporous framework MOF-74 analyzed using infrared spectroscopy," *Physical Review B*, vol. 81, no. 10, 2010.

[129] F. Bonino, S. Chavan, J. G. Vitillo, E. Groppo, G. Agostini, C. Lamberti, P. D. C. Dietzel, C. Prestipino, and S. Bordiga, "Local structure of cpo-27-ni metallorganic framework upon dehydration and coordination of no," *Chemistry of Materials*, vol. 20, no. 15, pp. 4957–4968, 2008.

[130] C. M. Brown, A. J. Ramirez-Cuesta, J. H. Her, P. S. Wheatley, and R. E. Morris, "Structure and spectroscopy of hydrogen adsorbed in a nickel metal-organic framework," *Chemical Physics*, vol. 427, pp. 3–8, 2013.

[131] L. Valenzano, J. G. Vitillo, S. Chavan, B. Civalleri, F. Bonino, S. Bordiga, and C. Lamberti, "Structure-activity relationships of simple molecules adsorbed on cpo-27-ni metal-organic framework: In situ experiments vs. theory," *Catalysis Today*, vol. 182, no. 1, pp. 67–79, 2012.

[132] H. Oh, *Doctoral thesis: Nanoporous Materials for Hydrogen Storage and H_2/D_2 Isotope Separation.* Universtät Stuttgart, Germany, 2014.

[133] W. Zhou, H. Wu, and T. Yildirim, "Enhanced H_2 Adsorption in Isostructural Metal-Organic Frameworks with Open Metal Sites: Strong Dependence of the Binding Strength on Metal Ions," *Journal of the American Chemical Society*, vol. 130, no. 46, pp. 15268–15269, 2008.

[134] H. Oh and M. Hirscher, "Quantum Sieving for Separation of Hydrogen Isotopes Using MOFs," *ChemPubSoc Europe DOI: 10.1002/ejic.201600253*, 2016.

[135] *Personal communication with D. Denysenko*. 2016.

Publications

- *Interplay of linker functionalization and hydrogen adsorption in the metal-organic framework MIL-101*
  P. A. Szilagyi, I. Weinrauch, H. Oh, M. Hirscher, J. Juan-Alcaniz, P. Serra-Crespo, M. de Respinis, B. J. Trzesniewski, F. Kapteijn, H. Geerlings, J. Gascon, B. Dam, A. Grzech, and R. van de Krol
  The Journal of Physical Chemistry C 118 (34), 19572-19579 (2014)

- *Formation of a quasi-solid structure by intercalated noble gas atoms in pores of Cu(I)-MFU-4l metal-organic framework*
  O. V. Magdysyuk, D. Denysenko, I. Weinrauch, D. Volkmer, M. Hirscher, and R. E. Dinnebier
  Chemical Communications 51 (4), 714-717 (2015)

- *D_2 and T_2 capture in the metal-organic framework Cu(I)-MFU-4l*
  I. Weinrauch*, I. Savchenko*, D. Denysenko, S.M. Souliou, H.-H. Kim, M. Le Tacon, L.L. Daemen, Y. Cheng, A. Mavrandonakis, A.J. Ramirez-Cuesta, D. Volkmer, G. Schütz, M. Hirscher, T. Heine
  Submitted to Nature Communications, July 2016

- *Fullerene cages encapsulated in a molecular framework material and quantitative gate tuning: C60@MFU-4*
  H. Oh*, I. Savchenko*, I. Weinrauch, A. Mavrandonakis, D. Denysenko*, M. Grzywa, D. Volkmer, T. Heine and M. Hirscher
  In preparation

- *Quantum Sieving of hydrogen isotopes in microporous metal-amide-imidazolate frameworks*
  I. Zaiser, S. S. Mondal, K. Behrens, D. Denysenko, D. Volkmer, H.J. Holdt, and M. Hirscher
  In preparation

Danksagung

An dieser Stelle möchte ich mich für die Unterstützung, die ich während der Erstellung dieser Dissertation bekommen habe, ganz herzlich bedanken. Insbesondere danke ich:

Prof. Dr. Gisela Schütz für die Möglichkeit am Max-Planck-Institut für Intelligente Systeme an meiner Promotion zu arbeiten, die Übernahme des Hauptberichts und Ihrer wertvollen Unterstützung während der Endphase.

Prof. Dr. Clemens Bechinger für die Übernahme des Mitberichts.

Dr. Michael Hirscher für die Korrektur und wissenschaftlichen Betreuung dieser Arbeit. Seine Tür stand für mich und meine Fragen immer offen.

Bernd Ludescher für seine kompetente Unterstützung bei allen technischen Angelegenheiten und seine Offenheit Neuem gegenüber, wie zum Beispiel den Vorbereitungen für das Raman Experiment.

Meinen Kollegen Dr. Julia Teufel, Prof. Dr. Hyunchul Oh, Rafael Balderas, Linda Zhang und Nadine Kotzur für die gute Zusammenarbeit und die tollen Gespräche.

Dr. Michael Haag, Dr. Christian Illg und Prof. Dr. Thomas Heine für interessante Gespräche auf dem Gebiet der Theorie.

Dr. Michaela-Sofia Souliou und Hun-ho Kim für die herzliche Zusammenarbeit und die intensive Betreuung beim Raman Experiment.

Dr. Timmy (A.J.) Ramirez-Cuesta und Dr. Yongqiang Cheng für die tolle wissenschaftliche Betreuung und die schöne Zeit am Oak Ridge National Laboratory.

Dr. Dmytro Denysenko, Dr. Phillip Schmieder, Sebastian Spirkl, Karsten Behrens, Dr. Laura McCormick, Philipp Müller, Prof. Dr. Dirk Volkmer, Prof. Dr. Hans-Jürgen Holdt, Prof. Dr. Pascal Dietzel und Prof. Dr. Stefan Kaskel für die Entwicklung und Bereitstellung der in dieser Arbeit untersuchten Materialien.

Therese Dragon, Monika Kotz und Ulrike Eigenthaler, die für Fragen und Angelegenheiten aller Art immer ein offenes Ohr hatten.

Meinem Mann Sebastian, für seine wissenschaftliche Unterstützung im Studium und während der Promotion. Insbesondere danke ich Ihm, dass er mich in turbulenten Zeiten aufgefangen hat.

Meinen Eltern Mihaela und Alfred, die mich mit Ihrer lieben und verständnisvollen Art mein ganzes Leben lang gefördert haben.

Eidesstattliche Erklärung

Hiermit erkläre ich an Eides statt, dass ich diese Arbeit ausschließlich unter Anleitung meiner wissenschaftlichen Lehrer und unter Verwendung der angegebenen Hilfsmittel angefertigt habe.

Stuttgart, den 25. August 2016

Ingrid Zaiser